SEX IN THE WORLD'S RELIGIONS

SEX IN THE WORLD'S RELIGIONS

GEOFFREY PARRINDER

GENERAL PUBLISHING CO. LIMITED

DON MILLS, ONTARIO

First published in Canada, 1980
by General Publishing Co. Limited
30 Lesmill Road, Don Mills, Ontario, Canada

Published simultaneously in Great Britain
by Sheldon Press
Marylebone Road, London NW1 4DU

Printed and bound in Great Britain by
Redwood Burn Limited
Trowbridge & Esher

ISBN 7736-1083-9

The Author

Geoffrey Parrinder is Emeritus Professor of the Comparative Study of Religions in the University of London. After ordination he spent twenty years teaching in West Africa and studying African religions, and became the founder member of the Department of Religious Studies in the University College of Ibadan, Nigeria. He has travelled widely in Africa, and in India, Pakistan, Sri Lanka, Burma, Iran, Israel, Jordan and Turkey and has lectured in Australia, America, India, Japan, and at Oxford. He is the author of many books on world religions which have been translated into ten languages.

Many waters cannot quench love
neither can the floods drown it:
if a man would give all the substance
of his house for love
it would utterly be contemned.

CONTENTS

		page
1.	INTRODUCTION	1
2.	SACRED SEX IN INDIA	5

Divine Examples. Heroic Ideals. Marriage and Ritual Union. Austerity and Chastity. Woman's Lot. Prostitution. Sex Manuals. Religion and Sex in Art. Yoga and Tantra. Reaction.

| 3. | BUDDHIST RENUNCIATION | 41 |

The Ascetic Middle Way. Celibate and Sexual Power. Buddhist Tantra. Lay Marriage and Morality.

| 4. | OTHER INDIAN TRADITIONS | 59 |

Jain Asceticism. Sikh Virility. Parsi Customs. Tribal Religions.

| 5. | CHINESE YIN AND YANG | 77 |

Female and Male. Yin and Yang. Tao. Tao in Sex. Confucian Morality. Marriage. Buddhist Influences. Variations. Reactions.

| 6. | JAPAN'S FLOATING WORLD | 103 |

Shinto Myth. Phallicism and Divine Unions. In and Yo. Women and Men. Marriage. Floating World and Geishas.

| 7. | TRADITIONAL AFRICA | 127 |

Attitudes. Myths. Phallicism. Initiation. Dowry and Polygamy. Fertility. Taboos. Change and Decay.

| 8. | ISLAMIC CUSTOMS | 151 |

The Prophet. Marriage in the Qur'an. Sex in the Traditions. Sex in Literature. Mystical Symbolism. The Status of Women: Early Islam, Veiling, Harems.

9.   HEBREW AFFIRMATIONS   178
     Creation. Phallicism and Circumcision. Male and
     Female. Love and Marriage. Symbolism.

10. CHRISTIAN DIVERSITY   202
     The Gentile Background. The Teaching and Practice of
     Jesus. Paul and Others. Monogamy and Love. The
     Ascetic Early and Medieval Church. Lay Marriage and
     Problems. Virgin Birth and Mother. Reformation
     Change and Conservatism. Modern Times.

11. MODERN INFLUENCES   240
     Modern Influences: Medicine, Psychology, Women's
     Rights, Religious Encounter.

     SELECT BIBLIOGRAPHY   249

     INDEX   255

Chapter 1

INTRODUCTION

Sex and religion are two of the commonest concerns of mankind. Often opposed as physical and spiritual, temporal and eternal, they seem to occupy different and clearly defined territories yet they are always crossing the frontiers. For man cannot live by bread alone and even his sexual intercourse is shot through with fantasy, while religion takes all the world as its province and turns its eyes upon the slightest manifestations of sex, as the history of the great world religions demonstrates.

But what is sex? The *Oxford English Dictionary* describes it first as 'Either of the two divisions of organic beings distinguished as male and female respectively; the males and females viewed collectively.' This is a very broad definition, including all men and women, and in this sense sex is fundamental to human nature. It will be important always to have this broad statement in mind, since it will be necessary to consider sex as involving the whole personality, of man and woman, and not to think of it as only sexual coition. Understanding of the role of man and the role of woman as full human beings is essential in the study of sex.

There is a further dictionary definition, however, which reflects change and restriction in the use of the word sex. 'In recent use often with more explicit notion: The sum of those differences in the structure and function of the reproductive organs on the ground of which beings are distinguished as male and female.' In modern times to speak of sex often does not mean simply gender, the differences of male and female, but more explicitly their physical union. Thus H. G. Wells,

1

writing in 1912 on *Marriage*, said that the young need to be told all we know of three fundamental things, 'the first of which is God . . . and the third Sex.' More recently people talk of 'having sex' or simply 'sex' as copulation.

Other related words have undergone similar changes. Thus 'intercourse' originally meant communication and in early use it was restricted to trade. In Victorian novels it was used of meetings and conversation, and their writers would have been shocked at its common modern equivalence with physical union, though technical writers in the nineteenth century already spoke of 'illicit intercourse'.

Because 'sex' and 'intercourse' have both wide and narrow meanings, the latter perhaps dominating nowadays, some writers have looked for other terms or qualifications. D. S. Bailey, an outstanding authority on the history of Christian sexual teaching, wrote of 'venereal intercourse' for what is technically called coition or sexual copulation. The word 'venereal' is derived from Venus, the Roman goddess of love, but unfortunately this connection is obscure in common English usage and 'venereal' too easily suggests an unpleasant disease communicated by sexual intercourse. It would be unfortunate to have this derogatory association of sex as dominant, especially in the study of religions that have traditions of depreciation of sex.

In this book efforts are made to be intelligible and accurate, though that is not easy. 'Sex' will be used in the wide sense of male and female relationships, and more particularly of 'sexual intercourse' in coition or copulation. The word 'love' will also be used in the broad meaning of loving relationships, and not merely restricted as is often done now to 'making love'.

'Religion' likewise is notoriously difficult to define. Some religions believe in a supreme God, others speak more of a power or truth, and yet others are chiefly concerned with the ancestors and survival of death. Taking the broad view again, the Chinese Yin and Yang will be included in the religious or spiritual, as well as the Almighty God or Word of the Lord.

This is a book of 'comparative religion', not in the sense of competition but in examination of different religious traditions. It seems to be unique in making a study of sex and religion together, as seen by the major living religions. For while there are countless books on the religions of the world, and summary comparative accounts of the major religions together, nearly all such works seem to neglect specifically sexual factors, despite the admitted importance of sex for religion. Many comparative studies of the Indian Upanishads, for example, expound their ideas of individual and universal souls, but they ignore Upanishadic teachings on rituals of sexual intercourse, and these are often omitted also in translations of the texts. Books on Yoga tend to leave aside sexual practices that may be used, or the physiological theories of sex that underlie much of both Indian and Chinese forms of Yoga.

Much of this work is exploratory, for relatively little has been written on such matters. In the Christian field, says Canon Bailey, 'no attempt has hitherto been made' to give a detailed account of the sexual tradition. On China, says Dr van Gulik, 'I found that there was practically no serious literature available, either in standard Chinese sources or in Western books and treatises on China.'[1] Some religions are better documented than others on sexual matters, and there is bound to be some imbalance in a comparative study. Apology is made here for such defects, and the hope is expressed that this important subject will be opened up more fully and in detail on particular religions in the future.

This book is not for Manichees, those who regard sexual intercourse as sinful, polluting, or inferior, and they are warned not to read it. If you can't stand the heat, get out of the kitchen. On the other hand, this book is not for the salacious, looking for saucy titbits. It is meant to be factual and scientific, and technical names have been used instead of their popular equivalents. The comparative aim has determined the contents, by trying to see others as they have related sex and religion to each other, and thereby perhaps to see ourselves better.

This study has aroused numerous reflections, not least on

---

[1] D. S. Bailey, *The Man–Woman Relation in Christian Thought*, 1959, p. vi; R. H. van Gulik, *Sexual Life in Ancient China*, 1961, p. xi.

the complex history of my own ancestral faith in this field. The activities of human beings often appear strange or foolish, but like other occupations the sexual life of mankind illustrates both the joy and the sadness that appear in human affairs. Some sexual practices hardly merit the name of love, yet love often emerges in unexpected places. Sometimes it is a will o' the wisp, and at other times it reflects Dante's vision of 'the love that moves the sun and the other stars'.

Chapter 2

SACRED SEX IN INDIA

———————

Nowhere have the close relationships of religion and sex been displayed more clearly than in India and, with divine and human models of sexual activity, sacramental views of sex were abundantly illustrated. It is helpful, therefore, to begin this comparative study with the ancient religion of Hinduism, though it is a vast complex wherein attitudes towards sex range from full indulgence to ascetic denial. Often regarded as the source of world-denying and pessimistic attitudes in Asia, Hinduism may also be seen as naturalistic and erotic. Some Hindu texts said that the sage should be indifferent to all human ties, and at the same time that it was the gods and sages who taught men the sciences of pleasure and love.

DIVINE EXAMPLES

Modern excavations in the ruined cities of the Indus plains take us back beyond the oldest scriptures, to find traces of sexual practices which link up with later times. According to the Aryan invaders of north-western India the peoples they conquered had black complexions, no noses to speak of, unintelligible speech, no rites, were 'indifferent to the gods', and probably worshipped the phallus.

Like most conquerors the Aryans despised their subjects because they did not speak their language or follow their religion. But worship of the phallus, the *linga* of later Hinduism, is apparent among the peoples of the Indus plains from many cone-shaped objects which have been found there

5

and are probably phallic representations. Among small engraved seals which have been found, several depict a horned figure, with three faces, sitting naked, in a Yoga-posture, surrounded by four wild animals. He is ithyphallic, and this with the animals and plant-like growth from his head indicates that he was a fertility god. This figure has been well called a Proto-Shiva, for in later Hinduism the great god Shiva was sometimes depicted with three or four faces, was a fertility deity, symbolized by the phallus, and known as Lord of Yoga and Lord of Beasts.

Also in the ruins of the Indus plains there have been dis-covered many rough terracotta statuettes of naked or nearly naked women, and these are thought to be icons of a Mother Goddess, very important for female fertility and the health of children. There are so many of these that they were probably kept in every home of the walled cities of the Indus valley. In the Aryan Vedic literature there is no great Mother Goddess, but if she went underground among the survivors of the Indus plains culture, she reappeared after a thousand years and it has been said that she became the greatest power in the Orient. Under many names—Great Goddess, Mother, Daughter of the Mountains (Parvati), Inaccessible (Durga), Black (Kali)—she is one of the great deities of modern India.

There is also an interesting bronze figurine which depicted a slim girl, naked except for necklace and bangles, and standing in a provocative posture. It has been suggested that she was a prototype of the *devadasi*, temple dancer and prostitute of later times, but it is not certain that she was a dancer or attached to a temple. There were also ring-shaped stones which have been claimed as representations of the vulva, the *yoni* revered in later Hinduism, but this iden-tification is disputed. There is another seal which shows a horned figure, perhaps a goddess, standing in a pipal tree, the sacred Bo-tree popular later. In front of her are seven pig-tailed figures, probably female attendants. Another seal shows a man fighting two tigers, which has been compared to the Mesopotamian motif of the hero Gilgamesh fighting two lions.

The cities of the Indus plains were devastated in the second millennium B.C. and there was no comparable architecture for nearly a thousand years. Religious ideas from the Indus plains cultures probably survived, and perhaps Yoga and the potent belief in rebirth may have come from those ancient times. They appear in the philosophical Upanishads, and rebirth at least is claimed there as a belief foreign to the Aryan priests. The god Shiva, unknown to the Vedic hymns, appeared in the Shvetashvatara Upanishad, where Yoga was also sketched. He was first of all shown as an attribute of the Vedic god Rudra, the name Shiva meaning 'kindly' or 'auspicious', but his mythology developed later.

In the great epic poem Mahabharata, Shiva appeared as one of the leading gods, and here and in the popular Purana tales his complex mythology and character were elaborated. Shiva was both a god of sex and of Yogic asceticism, of virility and destruction, and the *linga* by which he was represented showed his sexuality concentrated by asceticism. No one text gave all the episodes of Shiva's activities in sequence, but some of their main elements may be summarized.

Shiva was sitting in ascetic meditation in the mountains when Parvati, daughter of Himalaya, wanted to marry him. Her parents were displeased but Kama, god of love, was sent to inspire Shiva with desire for Parvati. Shiva burnt Kama to ashes with the fire of his third eye, but later revived him. Shiva then appeared before Parvati as a priest and tested her by describing his unpleasant qualities: his ashes, three eyes, nakedness, snakes, garland of skulls, and home in cremation grounds. When Parvati remained steadfast Shiva agreed to marry her, and at the wedding he caused scandal by appearing in scanty ascetic clothes.

After the wedding Shiva and Parvati made love, but following further adventures his sexual powers began to diminish and he went off to the forest to revive them by asceticism. The earth began to shake with his ascetic power (*tapas*), so that he abandoned it. He wandered in the forest, naked, ithyphallic, dancing and begging with a skull in his hand. The wives of the sages there fell in love with Shiva and followed him, so that the sages cursed his *linga* and made it fall to the ground. This caused a terrible fire and the gods Brahma and Vishnu tried in vain to find the top and bottom

7

of the huge *linga*, and peace was only restored when the sages and their wives agreed to worship it. When Shiva returned to Parvati the friction generated by their love-play was so great that the gods were again afraid and sent the god of fire, Agni, to interrupt them and eventually a son was born to the divine couple. [1]

These myths, with many variations, have been popular in India down the centuries and formed motifs for painting and sculpture. In many texts Shiva was said to be ithyphallic and this was his commonest iconic representation, with the *linga* often fixed in the female *yoni*. In the Linga Purana it was said that the gods went to visit Shiva in his paradise and found him in sexual intercourse with Parvati, which the couple continued despite their visitors. Vishnu began to laugh, and others were angry and cursed Shiva and his wife. They died in the position of intercourse, Shiva saying his new shape would be the *linga*, which men must model and worship. The *linga* was Shiva himself, and the *yoni* Parvati, and this was the origin of all things.

Sometimes Shiva was represented with his consort in a single androgynous human figure, as in a famous sculpture in the Elephanta caves near Bombay. The combination in Shiva of phallicism and asceticism may be confusing to foreigners, though to many Hindus they are not just opposites but interchangeable identities. Asceticism (*tapas*) and desire (*kama*) were not wholly contradictory, but were both seen as forms of energy. For many Hindus Shiva is the great God and his *linga*, often of carved stone, is the ubiquitous symbol of his cult. His modern worship, especially in southern India, has been called the clearest Indian form of monotheism, for devotees seek his favour alone and he is believed to appear in vision to inspire and help lonely souls. [2]

Hindu gods had their complementary goddesses, though most of them were pale reflections of their lords, having the same name with feminine endings, such as Indrani or Brahmani. The wife of the other great god, Vishnu, however,

[1] See W. D. O'Flaherty, *Asceticism and Eroticism in the Mythology of Śiva*, 1973, pp. 30 ff.

[2] See M. Dhavamony, *Love of God according to Śaiva Siddhānta*, 1971.

Shri or Lakshmi, was portrayed as a beautiful woman with a lotus in her hand, and she is popular as the goddess of good luck and worldly blessing. But it is the Mother Goddess, wife of Shiva under many names, who is the most important. She was the Shakti or potency of her male counterpart, active and immanent when he was passive and transcendent, and she was of great significance in later Tantra.

The Goddess was a great Mother, but she also had terrible aspects, and she is still represented in bazaar paintings as a horrible hag, black, and stark naked except for a garland of skulls. She has tusks and a lolling red tongue, brandishes weapons, and tramples on a buffalo-headed demon, or even stands on the prostrate body of her spouse Shiva. The Goddess is worshipped in the *yoni*, and according to myth when her father quarrelled with Shiva she threw herself into the flames of a sacrifice, and the ashes of her *yoni* fell in various places of India which became the shrines of her cult.

Another very popular Hindu god of love was Krishna, and he had an equally complex story. His name also means 'black', and it is possible that he was the successor of an ancient deity of dark-skinned peoples. He appeared in the Mahabharata as a hero of cowherds, though in the Bhagavad Gita he was the lofty teacher of morality and the supreme God. In Purana story Krishna was both cowherd adventurer and demon-killing prince. His childhood pranks made him the darling of women, and his romances with the cowgirls increased his attractions. When the girls were bathing, Krishna stole their clothes and made them come naked to him to receive them, with hands above their heads. This story of full frontal nudity, still popular in verse and painting, was given mystical interpretations of the nakedness of the soul before God.

Passionate longings were expressed in worship and poetry dedicated to Krishna. He played his flute in the forest and women left their husbands, breaking the customs which required women always to obey and stand by their husbands, but symbolizing the soul leaving all for God. The girls danced in a round, with Krishna in the midst, and by using his delusive power he provided each girl with a semblance of himself. They took off their clothes and jewels and offered them all to him, and after bathing with them he sent them home.

A theme that developed in myth was of Krishna disappearing from the dance to make love to one particular girl in the forest, and in time she was named as Radha his paramour, loveliest of all the cowgirls. The loves of Krishna and Radha became major themes, with her sexual passion and adultery in leaving her husband indicating the priority which God required in loving devotion. The Brahmavaivarta Purana, after the tenth century A.D., developed the eroticism of the cult and described the union of the couple in most sensual terms. 'Krishna pulled Radha with both his arms and stripped her of her clothes. Then he kissed her in four different ways, and the bells in her girdle were torn off in the battle of love. Then Radha mounted Krishna and had reversed intercourse, and later Krishna adopted eight different positions and tore her body with biting and scratching until she was unable to bear any more and they ceased from the battle.'

There were still the other cowgirls, now numbered as nine hundred thousand, and to satisfy them Krishna turned into an equal number of men so that each thought Krishna was loving her; they were beautiful in their nakedness and swooned in their pleasure until the park resounded with this mass intercourse. Such passionate language expressed the enthusiastic character of much Krishna worship, and to this day millions of pilgrims visit the sites of the myths and follow the themes of the story.[3]

HEROIC IDEALS

Hindu moral ideals for human relations were also expressed in the epic poems, Mahabharata and Ramayana and books of Sacred Law, almost as sacred as the Vedas and more loved. The adventures of men and women, rulers and heroes, gods and demons, in the epics played a great part in popular Indian religion and provided standards of conduct.

[3] See W. G. Archer, *The Loves of Krishna*, 1957; N. C. Chaudhuri, *Hinduism*, 1979, p. 275.

Four ideals or objectives of man were set out as basic concepts of Hindu life and conduct. These were: duty, gain, love, and salvation (*dharma*, *artha*, *kama*, *moksha*). In early texts they were usually the first three, since the last involved renunciation of other activities, in a search for liberation from worldly concerns. *Dharma*, duty or virtue, underlay all the other ideals, for material gain needed to be controlled by virtue, and love or pleasure should not be in conflict with the rights of other people.

These ideals were worked out for the basic classes of society: priests, rulers, merchants, and servants. The first three of these followed three or four stages of life: as students, householders, hermits, and ascetics. The life of the married householder was naturally the most common, and in some texts it was considered the greatest of the four stages. The student's life was preparatory, and the hermit stage or retirement should not come until the householder had seen his children's children, which assured a firm descent and performance of ancestral rituals. Asceticism was not for everybody and, for the householder, sex, like material gain, was not to be suppressed but rather regulated and developed, avoiding extremes of repression and licence.

In the Mahabharata, the longest poem in the world, Dharma is a constant refrain, and at the end it is declared that both gain and love flow from Dharma, so that it should be followed through pain and pleasure, since it is eternal. But while this moral ideal was upheld, the many tales of the epic depicted human successes and failures, love and cruelty, and divine and animal adventures as well.

The Bhagavad Gita was a small part of the Mahabharata, but it developed theological and moral ideas in a largely ascetic manner. Although the first word of the Gita was Dharma, the book was caught between affirming the need for action in the world and denial of worldly affections. The Gita did admit of its immanent God that 'where Dharma does not forbid it, in all creatures I am love (*kama*)'; and again, 'I am the generating god of love (Kandarpa)' (7,11; 10,28). But the prevailing tone of the Gita was passive or ascetic, though it rightly saw that renunciation did not lead to perfection by itself, that nobody could abstain from actions altogether,

11

and that ascetics might give up sensual actions yet still think about them, like St Antony in the desert.

In the Gita the warrior Arjuna was told to do his duty, but in complete detachment from results, without thought of reward or punishment. Moreover, he should see the same being in a priest as in an outcaste, and even more impassively should be the same to friend and foe, indifferent to enemy, neutral, or friend. There was little trace of human love or reference to women, and such ethereal teaching might be more appropriate to celibate priests or men in the last stages of life, rather than to ardent soldiers or married house-holders. This cold serenity has been criticized in modern times as disregarding the difference between good and evil, and giving the possibility of immoral actions. But it is not so harsh as one of the Upanishads, which declared that, 'He who understands me is not injured by any deed: not by stealing, not by procuring an abortion, not by the murder of his mother, not by the murder of his father.'[4]

Contrary to such indifference, the tales of the epic provide plenty of examples of affection, and the ideals of married love and the fidelity of women are popular themes. A favourite story of the Mahabharata is the love of Nala and Damayanti, of which there are several translations in English.[5] Nala was a king endowed with all the virtues, handsome, a connoisseur of horses, who spoke the truth but loved to gamble. Damayanti was exceedingly beautiful, with long eyes and flawless limbs, and she fell in love with Nala and he with her, simply by hearing each other's praises sung. Damayanti's father decided to hold a 'self-choice' or 'bridegroom-choice', a privilege of daughters of the warrior caste. Four of the chief gods attended and each took the form of Nala to confuse the girl, but she was able to distinguish her beloved by his shadow, faded garland, dust, sweat and blinking eyes, none of which marked the perfect gods. Nala and Damayanti were married and lived in happy love until Nala lost all his belongings in gambling, like some other heroes, so that he had to go to the forest with his wife.

[4] Kaushitaki Upanishad, 3.1
[5] Mahabharata, 3,50 ff; see translation by J. A. B. van Buitenen, vol. 2, p. 322 ff.

In despair he abandoned Damayanti in the hope that she would return to her father, and being bitten by a snake Nala was turned into a dwarf. After many adventures the couple came together again, when Damayanti recognized Nala and he regained his own form. He won back his kingdom at dice and they lived in happiness, Nala content in reunion with his wife and Damayanti refreshed 'like an acre with half-grown crops on receiving rain'. He who tells or listens to this story will find sons and grandsons, and be happy in health and love.

A faithful and sacrificial lover was princess Savitri, named after a goddess. Although she was beautiful, with fine waist and broad hips, like a golden statue, yet no man asked for her hand, being dazzled by her eyes like lotus petals which blazed with splendour. So her father gave her a 'self-choice' and Savitri rode off in a chariot to inspect the field. She chose Satyavat, a prince who had only a year to live, and insisted on marrying him despite her father's objections. When the time came for his death Satyavat laid his head in Savitri's lap, and she saw Yama the god of death appear and draw out Satyavat's soul, the size of a thumb. Savitri followed Yama and he was pleased by her devotion and offered her three boons for her family, except the life of her husband, but she persisted and gained that also. Satyavat woke up and they returned to his parents, where the story was recounted. Thus Savitri by wifely devotion saved her husband, his family and the entire dynasty. Satyavat received a life of four hundred years, and Savitri was praised for rescuing him and she gave birth to a hundred sons who increased her fame.[6]

Generally the popular tales give great importance to Kama: love, pleasure, or desire. Despite what was said earlier about Dharma, other passages in the epic speak of Kama as the foundation of Dharma and Artha, it is their essence and womb, and the innermost core of the world. Without love a man has no wish for worldly profit, and all kinds of occupations are inspired by it. For the sake of love sages studied the Vedas, offered sacrifices, or gave themselves

[6] Mahabharata, 3.277 ff.

13

up to asceticism. Love is all-powerful, knows no laws, and overwhelms with passion. But love must be on both sides, and lead to the pleasures of sexual union, for if a man loves a woman who does not love him then his whole body glows in torment. Love therefore should be enjoyed with discretion, and be mutual.[7]

The other epic poem, the Ramayana, has always been taken as one of the most inspiring examples of heroic courage, virtue, and marital fidelity. This 'story of Rama' told of the rightful heir to the throne of Ayodhya being deprived of the succession, and retiring to the forest with his wife Sita and his brother for fourteen years. There they led a hard life, sleeping on the ground and clothed in bark, while sages came to ask their help against demons that were afflicting them. Rama fought the demons, whereupon their king Ravana abducted Sita and took her to his island fortress in Shri Lanka (Ceylon). With the help of the monkeys Sita was rescued, but she had to undergo an ordeal to prove her chastity and fidelity to Rama. She emerged victorious and the pair were restored to the throne, there to reign in righteousness.

The story is straightforward, and little emerges of the personal relationships of Rama and Sita. But in a later section Rama was still suspicious of Sita's conduct during her captivity and he proposed another purification. Whereupon Sita exclaimed, 'I'm going home to mother', and the earth swallowed her up. Sita, whose name meant 'furrow', was the daughter of Mother Earth and returned to her true home. Then Rama entered a river, like the travellers in *Pilgrim's Progress*, and 'all the trumpets sounded for him on the other side' when with heavenly music he returned to the god Vishnu. For in devotion this heroic story developed a religious theme of the avatar or 'incarnation' of Vishnu in Rama. Rama and Sita became both marital and religious ideals, and in devotional movements which swept across India drama and song, temples and sculptures, were dedicated to them and made them popular with the masses. Rama and Sita are celebrated in annual festivals, plays of

[7] Ibid., 12.167.

their story attract crowds and have spread across south-east Asia, while the cinema gives lush versions of their adventures and love.

MARRIAGE AND RITUAL UNION

Rituals of sexual union can be traced back to the Vedas, though these were chiefly concerned with the worship of nature gods and details came later. The hymns of the Vedas were mainly invocations of these gods and were meant to accompany a libation, Soma, and a fire sacrifice of melted butter. In a hymn to Heaven and Earth the divine fire, Agni, was said to 'milk from the bull abounding in seed his shining moisture', meaning that Agni by his fire caused heaven to fertilize the earth and the latter to be productive. [8]

An ancient Vedic ceremony was the 'horse-sacrifice', which was the greatest animal sacrifice of those times. After the king had let a horse roam at will for a year, to enlarge his territory, it would be sacrificed and his wife lay by the horse and imitated copulation with it. In the Mahabharata King Yudhishthira let loose a horse in this way before his enthronement, and at the eventual sacrifice Draupadi, the common wife of the five Pandu princes, lay by the horse.

Marriage ceremonies came to be the most elaborate of domestic sacrifices, though only a few of the later rituals can be traced back to the Vedic period. Marriage was regarded as a sacrifice in itself, and an unmarried man was called 'one without a sacrifice'. Priests as well as laymen were married, and where there was a vocation to celibacy it was generally reserved for ascetics and was not necessarily lifelong. A Brahmana text said that 'he who has no wife is without a sacrifice', and added that 'he is himself a half man and the second half is wife.' Men yearned for sons, though occasionally it was said that sages had passed beyond such desire. [9]

In olden days marriage ceremonies varied, though they were arranged by the fathers and celebrated on auspicious

[8] Rig Veda, 1.160.
[9] J. J. Meyer, *Sexual Life in Ancient India*, 1952, p. 150 ff.

days. The father would deck his daughter with clothes and jewels, and the bridgegroom might give as much wealth as he could afford to the bride and her kinsmen. The father addressed the couple with the simple words: 'May both of you perform together your duties.' But more detail appears from the epics where the parents gave many presents to the family, the dead, and holy men. At Sita's wedding to Rama, her father lit the flame on the family altar and sacrificed with traditional verses. Then he placed the adorned Sita before the fire, facing Rama, and said: 'This is Sita, my daughter, your wife. Take her, I beg; take her hand with your hand. As a faithful wife, favoured by happiness, she will follow you evermore as your shadow.'[10]

There are many ceremonies in modern traditional marriages, for those who can afford them, and in many of them the bride and bridegroom represent the god Shiva and his wife Parvati, personifying them as ascetics but also as splendid deities. The bridegroom wears a loincloth but also a crown covered with gold or silver paper. Other couples might be more fully dressed, and in great procession the bridegroom comes on a horse, perhaps borrowed for the occasion, and in royal weddings he would be mounted on an elephant. Those who are interested in the many details of marriage ceremonies may be referred to specialist works on them.[11]

The custom of child marriage goes back to ancient times, though apologists claim that the laws were given only as guidance to suitable ages. The principle seemed to be that the husband should be three times the age of his wife: 'A man of thirty years shall marry a maiden of twelve, or a man of twenty-four a girl of eight; but if the performance of his duties would otherwise be impeded, he must marry sooner, for the husband receives his wife from the gods and not according to his own will.'[12] The rules of caste hedged around the options open in the choice of a wife, though the man and her father had the responsibility and not the woman.

[10] Laws of Manu, 3.26 ff.; Ramayana, 1.73.
[11] S. Stevenson, *The Rites of the Twice-born*, 1920, chs. 3–5; R. B. Pandey, *Hindu Samskāras*, 2nd edn. 1969, ch. 8.
[12] Laws of Manu, 9.94.

Similarly the remarriage of a widow was forbidden from an early date, for she must always revere him. 'A husband must be constantly worshipped as a god by a faithful wife, even if he is devoid of good qualities or seeks pleasure elsewhere . . . And a virtuous wife who remains constantly chaste after the death of her husband, reaches heaven, but a woman who from a desire to have offspring violates her duty towards her deceased husband, brings disgrace on herself and loses her place with her husband in heaven.' Observers in modern times have noted that 'the horror of being left a widow gives colour and direction to all a Hindu woman's prayers and thoughts.'[13] Pleasure in sex may be very limited, for a woman.

Ancient texts gave many love-charms for a successful family life, and to win or compel a woman's or a man's love. There were imprecations against rival women and curses of spinsterhood. A lover entering a girl's home by night muttered a spell to put the household to sleep. Special attention was given to maintaining or recovering virile power in the man, and successful conception and pregnancy in the woman, with protection against demons and eventual birth of sons. There were medical and magical remedies, herbalism, and charms.[14]

In the Upanishads, alongside philosophical dialogues, there are descriptions of ritual sexual intercourse. In an early passage of cosmological speculation it was said that in the beginning there existed only the self in the form of a person. He was alone and afraid, and had no pleasure, so that he desired a second. This self was as big as a man and woman in close embrace, so he divided himself into two parts from which husband and wife arose, for 'oneself is like a half-fragment', therefore this space is filled by a wife. Then he copulated with her and human beings were produced. She hid herself and became a cow, but he changed into a bull and from their copulation cattle were born. Likewise with all other animals, he created all, whatever pairs there are, even down to the ants. Whatever is moist, he created from semen, and from the mouth as fire-hole or vulva (*yoni*) he created

[13] Ibid., 5.154 ff, and S. Stevenson, op. cit., p. 108.
[14] See H. Zimmer, *Philosophies of India*, 1951, pp. 147 ff.

fire.[15] Plato, in his *Symposium*, described a similar single original being which was separated into man and woman.

In later verses of the first Upanishad sexual intercourse was described as a ceremony, with preliminary purifications, symbolical comparisons, and prayers, as in other Vedic rituals. The example was given of Prajapati, Lord of Creatures, who created woman and revered her lower parts, which his followers should do. Then he enlarged that stone which projects and impregnated her. The woman was transfigured to become the consecrated place where sacrifice was performed. 'Her lap is a sacrificial altar, her pubic hairs are the sacrificial grass, her skin is the press for the Soma libation, the two lips of the vulva are the fire in the middle.' Indeed, just as one is strengthened by sacrifice, so great is the world of him who practises sexual intercourse with this ritual knowledge. Thus coition was not a hurried, or only a physical, incident but an engagement of the whole person in a sacrament.[16]

From this time, at least, there spread the belief that the rewards of a sacrifice could be obtained by a ritually consummated sexual union. But the male was dominant, for if a man desired a woman then, after her purification from menstruation, he would invite her to lie with him. If she did not grant his desire (*kama*) he would bribe her, and if she still refused he should hit her with his hand or with a stick and overcome her and say, 'With power, with glory I take away your glory' and she became inglorious.[17]

Belief in the mystical power of semen, which was important in later Yoga, appeared here. If semen was spilt, whether awake or asleep, it should be taken between finger and thumb and rubbed between the breasts or eyebrows while saying, 'I reclaim this semen, let me come to strength again'. Magical notions of the reabsorption of semen appeared in instructions for behaviour according to whether offspring was desired or not, a kind of contraception by supernatural means. If a man desired a woman with the thought, 'may she not conceive offspring', then after inserting his sexual organ (*artha*) in her, and joining mouth to

[15] Brihad-aranyaka Upanishad, 1.4.   [16] Ibid., 6.4.   [17] Ibid.

mouth, he should breathe in and out and says, 'With power, with semen, I reclaim the semen from you.' But if he wished the woman to conceive he would say, 'With power, with semen I deposit semen in you.' If a man's wife had a lover the husband would cast a magical charm on him, putting a row of arrows in reverse order and sacrificing in reverse order and saying, 'You have made a libation in my fire, I take away your breathing in and out', and naming the enemy.

Conception took place in the name of the gods when a man desired a son. In the morning libations were made to the gods, and having eaten and washed the man sprinkled the woman three times with water saying, 'I am this man, you are that woman . . . I am heaven, you are earth.' He opened her thighs saying, 'Spread yourselves apart, heaven and earth'. He inserted his sexual organ in her, and joining mouth to mouth he stroked her hair three times, saying, 'Let Vishnu make the womb prepared . . . Prajapati, let him pour in', and ending,

> As Earth contains the germ of Fire,
> as Heaven is pregnant with the Storm,
> as the Wind is germ of the Directions,
> even so I place a germ in you.[18]

Sexual intercourse was not only a physical action, it was given the value of a religious ritual, and thus prepared the way for the later developments known as Tantra. In corresponding fashion, ritual could be interpreted in a sexual way and even minute details might be explained by this symbolism. Thus if in the course of a recitation the priest separated the first two quarters of a verse and brought the other two close together, this was said to be happening because a woman separates her thighs during copulation and the man presses them together. The inaudible recitation of a text was compared to the emission of semen, and when a priest turned his back and went down on his knees this was explained by the imagery of the copulation of animals.[19]

In the second of the classical Upanishads sexual union was

---

[18] Brihad-aranyaka Upanishad, 6.4,22.   [19] Aitareya Brahmana, 10.3.

transposed and valued as a liturgical chant (*saman*), and in particular the Vamadevya, the melody which accompanied the pressing of the plant for the midday libation. Each of the six actions of intercourse was made to correspond with parts of the ritual: the summons, the request, lying down with the woman, lying upon her, coming to the end, coming to finish. The conclusion was reached that, 'He who knows thus this Vamadevya Saman as woven upon copulation comes to copulation, procreates himself from every copulation, reaches a full length of life, lives long, becomes great in offspring and in cattle, great in fame. One should never abstain from any woman. That is his rule.'[20]

In the epic it was said that the adult state was reached when people 'become ripe for love', and married couples must perform their sexual duty. The term *ritu*, a time or season, was used for both menstruation and for the following days which were favourable for procreation. It was a sin for a husband not to visit his wife at the latter time, though he must never approach her during her periods, except in some forms of Tantra when this was encouraged by breaking the taboo. Wise men should go to their wives at the latter *ritu*, and keep away from strange women, and this was said to be an ancient and sacred rule for all four castes which was followed in the golden age.

Sexual intercourse was not unrestricted. It should be practised privately and not in the open air, and only with the vulva since oral sex was forbidden. Intercourse should not be held with strange women, and especially not with those of baser caste, with exceptions again in Tantra. Both the epics and the Laws of Manu condemned intercourse with a teacher's wife as particularly shameful, though since the prohibitions detailed anointing her, helping her in the bath, shampooing her limbs, and arranging her hair, the dangers must have been great for young pupils, and prohibition implies temptation. The same Laws forbade rape, to be punished by severing two fingers, and Lesbianism which brought fines and beating to a girl, and head shaved or two fingers severed for a woman who polluted a girl and she had

[20] Chandogya Upanishad, 2.13.

20

to ride through the town on a donkey. Male homosexuality seemed to receive less punishment, 'a twice-born man who commits an unnatural offence with a male shall bathe, dressed in his clothes', says one law, but another prescribed loss of his caste. Some medieval writers regarded 'under-love', male homosexuality, as quite common and not a perversion.

Incest was severely punished, including sexual intercourse with sisters by the same mother, wives of a friend or of a son. This was equivalent to violation of a Guru's bed, for which punishments varied. They could involve lying on a heated iron bed or embracing the red-hot image of a woman, or cutting off penis and testicles and walking with them in joined hands until the offender fell down dead. Or he could do penance in the forest for a year, or live on barley gruel for three months and perform penance. [21]

AUSTERITY AND CHASTITY

Despite general positive attitudes to sex, some found all worldly affairs distasteful, and there developed renunciation of normal life by Hindu ascetics and Jain and Buddhist monks. In the Upanishads King Brihad-ratha, having established his son in the kingdom, became indifferent to the world and went into the forest to stand, with arms erect and looking at the sun, for a thousand days. To a visitor he said, 'In this ill-smelling, unsubstantial body, which is a conglomerate of bone, skin, muscle, marrow, flesh, semen, blood, mucus, tears, rheum, faeces, urine, wind, bile, and phlegm, what is the good of enjoyment of desires?' [22]

A dualism developed, such as was found in other religions, between body and soul. There was no suggestion that sexual intercourse might have mystical meanings and actually help in attaining salvation, but on the contrary the suppression of sexual and all desires was seen as a condition for liberation. Similarly, in the epics millions of sages were said to have been wholly continent and thus to have burnt up all their

[21] Laws of Manu, 8.367 f.; 11. 59.104 f., 175.
[22] Maitri Upanishad, 1.3.

21

sins. Yet their very asceticism, being a powerful force, became dangerous to the stability of the world and threatened the very gods. Nymphs were sent to tempt such ascetics and they promptly lost their devotion by seeing beautiful women.

When the god Indra saw the danger of the powerful austerities of the seer Vishvamitra he sent the nymph Menaka to distract him. The girl, with pretty waist and lovely buttocks, greeted the ascetic and began to play in front of him. The wind blew off her skirt so that she appeared naked, whereupon he lusted to lie with her and they made love in the woods for a long time, which seemed only a day. On Menaka the hermit begot Shakuntala, a heroine of later story.

There are numerous stories of ascetics who had an involuntary orgasm at the sight of a lovely woman. Indra sent another divine nymph to check the austerities of Sharadvat. When he saw her in the forest wearing only a single cloth, with a figure unparalleled in the world, the sage stared at her and shuddered. He held his ground but a sudden spasm overcame him that made him spill his seed without noticing, and it fell into a reed stalk which split in two and from it twins were born.

The warrior Arjuna, however, was not so easily seduced. Indra sent the nymph Urvashi to tempt him, and her charms were lusciously described: the moon of her face, hair full of jasmine flowers, black-nippled breasts shaking up and down, temple of the god of love like a mountain, ankles hung with bells, and long red toes. But Arjuna was not moved, he bowed to her as a servant, stopped his ears at her words and honoured her as his mother, so that Urvashi cursed him with impotence. Yet Arjuna had had many wives and mistresses, and this story showed Indra approving the resistance of his son to temptation, though the curse of impotence remained for a year. In the last book of the Ramayana, the same nymph Urvashi was seen sporting in the waters by the god Varuna and he lusted after her, but she claimed to be already bespoken so that he directed his seed into a pitcher.[23]

[23] Mahabharata, 1.66; 1.120; 3.45; Ramayana, 7.56.

The epics often declared that chastity is the highest virtue. Men should not listen to light talk about women or look at them unclothed, and if a man was inflamed with improper desire he should put himself into water to cool off. The chastity of women was even more important. The polyandrous Draupadi 'each day became a virgin again' from her five successive unions until the last, so that each of the brothers had her untouched. Modesty was insisted upon, and both bathing and sleeping naked were forbidden, with exceptions in the stories of Krishna. It is strange that, despite the delight of Indian literature in sensual charm, the nude rarely appeared in Indian painting, and many statues with their full portrayal of the rounded female body yet suggest that it is clothed in a fine garment fastened at the wrists and ankles.

WOMAN'S LOT

In heroic times the epics depicted women desiring sexual union as avidly as men, and sometimes they might be even more erotic. The sage Agastya formed a superb woman from parts of different creatures, had her born to a king who was pining for a child, and when she was nubile he married her. The wife, Lopamudra, with long eyes and thighs like plantain stems, discarded her fine clothes and put on rags at her husband's bidding. But when she had passed through menstruation and had bathed, and was luminous with beauty, Agastya summoned her to intercourse. However Lopamudra, with folded hands and blushing, and with love-pleading words, asked the sage not only to take her for the sake of children but to give her as much pleasure as he found in her. He should lie with her on a fine bed, deck her with jewels, and approach her in garlands and ornaments. This set the husband on a search for wealth and adornment, and finally he had to do everything that his wife wanted. She said, 'you have done my every desire', and the hermit faithfully lay with his wife who equalled him in virtue.[24]

The nobility and dignity of woman is said to be fair, holy,

[24] Mahabharata, 3.95 f.

and according to Dharma. A faithful woman honoured her husband above ascetics, she could work miracles, and had her reward in the world to come. But the husband's demands were stern, she should follow his law alone, and when some husbands were too harsh their wives found it hard to bear. Women were honoured on the one hand, and on the other they could be accused of falsehood, trickery, unchastity, and being the essence of evil. Women were treated as chattels when girls were given as gifts, daughters sacrificed to save the father, wives or daughters offered to satisfy the lusts of guests.

The Laws of Manu said that 'Day and night women must be kept in dependence by the males of their families, and if they attach themselves to sensual enjoyments they must be kept under one's control.' A woman was protected by her father in childhood, her husband in youth, and her sons in old age, but she was 'never fit for independence'. Women should be guarded against evil inclinations, and employed in household, religious, and commercial duties. They should be kept from alcohol, sleeping at unseasonable hours, rambling abroad, and other men's houses. Many faults are attributed to them, in texts clearly compiled by men.[25]

The woman's hope was to bear children, since the mother was the centre of the family. She was both the physical and spiritual teacher of her children, and many texts say that 'the mother is the highest guru', 'she stands above ten fathers', and 'there is no higher Dharma than truth and no guru to equal the mother'. Maternal love was a constant epic theme, and children both revered and loved their mother. It followed that when children left home, or the husband died, a woman's lot was hard and it was said that 'widowhood is the greatest sorrow'.

Women were often said to have declared that they wanted to follow their husband in death, but the widow-burning of later times and particular regions seems to have had little basis in classical days. There is only one sure instance in the many tales of the Mahabharata and one in the late seventh book of the Ramayana. *Sati* (suttee), 'faithful', widow-

[25] Laws of Manu, 9.

burning was known in India before the Christian era, however, and commented on by the Greeks. It developed in the early centuries of our present era and spread to southern India by the tenth century, when it is said that sometimes thousands of a king's wives were burnt with his corpse. A few Hindu lawgivers and poets condemned the practice, but its abolition had to wait till modern times under nineteenth-century British imperial rule, though occasional instances have been reported in this century.

There is no doubt, despite some denials, that countless widows were burnt with their husbands on the funeral pyre, sometimes willingly and sometimes forcibly. There are many reports by external observers, ranging from the eighteenth to the twentieth centuries, and details may be found in particular studies and legal documents.[26]

The veiling and seclusion of women were particular Islamic customs, and will be considered under that religion. But the Muhammadan invasions of India, from the eleventh century intensified restrictions upon women that were already present by Brahminical laws and customs. In Hindu society it was the man, father or husband, who had complete disposal of the woman, daughter or wife. To this day many men, especially Brahmins, are waited on hand and foot by their wives. According to tradition a Brahmin wife should rise first to wash herself and clean the house; only then should she wake her husband, standing at a distance from his bed (since they sleep in separate beds, and in separate rooms after the first child is born). She bows with folded hands, saying 'Hail, Lord Krishna'. If she had time she might worship his right big toe, marking it with sandalwood and offering incense and lights as to a god.

However, sixty years ago a Brahmin lady who checked what Mrs Stevenson was recording on these duties, said laughingly: 'That is the way, no doubt, that we ought to pay our reverence to our husbands, but we have not time nowadays . . . In the mornings all I have time to do is to stand at the bottom of his bed and say: "Utha-Utha!" (up you get!), and after that I am far too busy cooking for him to have any

[26] See E. Thompson, *Suttee*, 1928.

25

time to waste in worshipping him!'[27] If that was so in 1920, it is likely to be an even more widespread attitude today after the great part that women have played in Indian social and political movements. However, older women and village women may keep some of the traditional customs, and there is sometimes revival of the old and bad, as well as the old and good, under nationalistic resurgence.

PROSTITUTION

In the epics and classical Indian texts extra-marital sex, for men, was accepted as part of life. There were high-class courtesans for male needs, and they were expected to be educated in sixty-four arts and sciences and were the equivalent of Greek *hetairae* or Japanese geishas. The courtesans were women of high education, whose teachers should be paid by the state, and their accomplishments included dancing, singing, acting, sewing, flower-arranging, and other useful and domestic arts. The cultured courtesans (*ganikas*) were distinguished from low-class and promiscuous prostitutes (*kalutas*).

The contrasts of traditional Indian life were illustrated in the opposition of the ascetic and the courtesan, or the married heroine and the prostitute. Yudhishthira, the righteous king of the epic, not only had his wives and a share in the fair Draupadi, but he sent greetings to the 'fair-clad, ornamented, scented, pleasing and happy women of the houses of joy'. Accompanying Rama's army in search of his faithful wife Sita, were 'women that live by their beauty', and these were not merely camp prostitutes but women prominent in the life and affairs of towns. When Rama was to be consecrated as king his priest directed that everything should have a festal air, the temples be put in order and the daughters of pleasure be arranged in the royal palace.

On the other hand, revealing tensions of Indian life and teaching, the epics also contained attacks on ordinary prostitution. Brothels and drinking halls were to be held in check as harmful to the kingdom, and a harlot was said to be

---

[27] S. Stevenson, *The Rites of the Twice-born*, p. 251.

a hundred times worse than a slaughter-house which destroyed living beings. Thieves and criminals were associated with prostitutes, and the Kama Sutra said that courtesans should be attached to police or powerful men to protect them from being bullied.

The *deva-dasis*, 'god's servants', were high-class prostitutes attached to the service of deities in Hindu temples. This practice went back to ancient times and developed until the end of the last century. Girls were given to temples in childhood as a gift to the god, perhaps in the hope of getting a son or some other benefit. They were said to be married to the deity, often Krishna or Shiva, and in a formal wedding ceremony they might be ritually deflowered by a priest or rich patron, or made to sit on a stone *linga*. The girls were trained in erotic arts and made available to temple visitors, for a price. Because their duties included dancing and singing, the dance was often regarded as immoral until modern reforms tried to purify it.

Temple harlotry in India became notorious, and pilgrims sometimes complained that they were hindered in worship by the seductions of the temple girls. Large temples, especially in south India, often seemed like brothels to outside observers, with hundreds of prostitutes who were taxed by the local states. Perhaps from this example of taxation, lay prostitution was exploited as a source of private and public income.

Temple prostitution was closed by the British, with the help of Hindu reformers, along with the suppression of widow-burning. Earlier, under Muslim rule, professional courtesans who were not attached to Hindu temples specialized in erotic dances and were very popular. These were the 'nautch girls' of India and southern Asia, a name that was derived from the Hindi *nach*, meaning 'to dance'. In the present century efforts have been made to free dances from erotic associations, and the classical styles have been revived, by male dancers and also by women. Many temples have preserved traditional dances, often only performed by men with the female parts being taken by boys, and the performers wearing masks to symbolize the gods or their attendants.

SEX MANUALS

The sexual arts were said to have been first promulgated by the Lord of Creatures, Prajapati, and codified in a long treatise which took the example of the amorous play of the divine Shiva and Parvati. This mythical treatise was supposed to be the basis for numerous works on eroticism of which only fragments survive, discussing courtship, sexual union, married love, prostitution, texts (*mantras*), spells, potions, aphrodisiacs, and special appliances. Later the intercourse of Krishna and his paramour Radha were the subject of many erotic writings and paintings.

The most justly celebrated Indian erotic work, from which many later writers borrowed, was the Kama Sutra, 'Love Text'. This was attributed to a Brahmin priest, Vatsyayana, in the third or fourth century A.D., and according to tradition he remained a lifelong celibate and ascetic, though it is hard to believe that he did not have the backing of some personal experience. The book is a masterly but abbreviated version of materials from earlier tradition, and it served as a pattern for generations. It is primarily the book of the householder, with instructions on sexual techniques, without mystical ideas of union with the deity through sex which may have been in the older tradition and which appeared in later Tantra.

The Kama Sutra opened with the praise of Dharma, Artha, and Kama, which had been regulated in commandments given by the Lord of Creatures in ancient texts. The latter two, gain and love, should be studied in youth and Dharma in old age in order to obtain salvation. Kama is the enjoyment of appropriate objects by the five senses, assisted by the mind and the soul. To the objection that sexual intercourse is practised even by animals and does not need any work on the subject, the answer is given that it depends on men and women, requires the application of proper means, and differently from animals it requires thought, for practice at any season.

The Kama Sutra maintains that women as well as men should study its arts, girls as a preparation for marriage and afterwards with the consent of their husbands. It lists the sixty-four arts, mostly household but including reading, poetry, and making mystical diagrams. For men there was skill in sports, arms, and gambling. Courtesans would be versed in these arts, but married women would increase their attractiveness thereby, and find means of support if they were separated from their husbands. Other teachings discuss the arrangements of a house, the daily life of citizens, and the kinds of women and friends to be cultivated.

The central part of the Kama Sutra gives details of practices in sexual intercourse, distinguishing between suitable and unsuitable partners, different kinds of embraces, and varieties of kissing, pressing, scratching, and biting. Women of different regions should be approached according to their own customs, and in 'equal', 'high', or 'low' intercourse the partners should lie in the most pleasurable manner. Details are given of various positions, some of which are acrobatic or require much practice, and male dominance appears when a man 'enjoys two women at the same time' or many women together. Women may act the part of men, when their partners are tired or for variety, but oral sex is reserved for eunuchs disguised as males or females, and Vatsyayana said that it should never be done by a learned Brahmin or by 'a man of good reputation', because although it was allowed by the texts there was no reason for the practice, or only exceptionally.

Other parts of the Kama Sutra discuss courtship, by creating confidence in the girl and by a girl seeking to gain over a man, betrothal, and forms of marriage. Then the behaviour of a virtuous woman is discussed, and her conduct during her husband's absence, but not his. The women of a king's harem should attend to his needs, and a polygamous man should act fairly towards all his wives. Women should reject the advances of other men than their husbands, though men seek to gain other women, and a further section discusses courtesans seeking out men, getting money, and the way to get rid of a weary lover. Concluding remarks discuss personal adornment, the enlargement of sexual organs, love philtres, and magical potions. The Kama Sutra was composed for the benefit of the world, while the author was studying religion and engaged in the contemplation of the deity.

29

The influence of the Kama Sutra and similar works upon sculpture, painting, and literature was of great social importance, before the puritanism of recent times sought to cover all sex from public view or discussion. For nearly a thousand years the Kama Sutra was the standard work on sex, both coition and other relationships of men and women. In ancient Indian society the sexes could mingle more freely than they did later, there was pre-marital and extra-marital sexual relationship, and while marriage was normal it was often polygamous, for men who could afford it. Women were not secluded, as they were in zenanas and harems under Muslim rule, although the later seclusion produced its own eroticism, with advantages to the male.

About the twelfth century Koka or Kokkoka set out to expound the pleasures of sex from his own experience, in the Koka Shastra or Ratirahasya, the 'Secret Doctrine of Love's Delight'. According to legend a nymphomaniac appeared at the court of Koka's patron, threw off her clothes, and declared that since neither gods, demons, or men could satisfy her she would wander the world naked until she met her match. Koka joined his hands and asked permission humbly to tame the shrew, which he did so forcibly that she fainted from repeated orgasms and was rid of desire for her next seven incarnations. This Shastra is a handbook of love-making, discussing different types of women by their genital characters, customs, and temperaments; courtship, embraces, kisses, love-spells, and more detail than the Kama Sutra on coition and postures of intercourse.

In the sixteenth century the Ananga Ranga, 'the Stage of the Love God', became one of the most influential erotic texts by showing how a married couple 'may pass through life in union'. Sex was said to be disastrous outside marriage, but since the cause of such diversion was monotony and 'want of varied pleasures', the author showed how husband and wife might live as with thirty-two partners by varying into as many positions of intercourse, so that satiety became impossible. Physical enjoyment was essential, and with it husband and wife could live together 'as one soul in a single body', which would ensure happiness in this life and in the world to come. Other unions were strongly forbidden: to look on a Brahmin's wife with desire was improper and to lie with her was deadly sin. Even legitimate times and places were restricted: sexual intercourse should only take place at night, in certain places, and in mild weather, though the varieties of love-making that were recommended would make up for these restrictions in some degree. All these manuals of sex taught that care, time and detail were necessary for sexual relationships in general and intercourse in particular. It may be understood that the hurried copulations of Europeans in India caused them to be called 'dung-hill cocks', for the physical was isolated from the personal and spiritual.

RELIGION AND SEX IN ART

The influence of the Kama Sutra and other manuals was felt in many forms of literature and even in devotional poetry. This was not surprising, given the traditional sacramental view of sex, the profound awe in which coition was held, through which the Creator continued his work in mankind. The models of divine lovers, Shiva and Parvati, Krishna and Radha, served both for devotion and for inspiration of sexual union.

In the twelfth century the Gita Govinda, 'Song of the Cowherd', was written in Sanskrit by the Bengali poet Jayadeva. The poem described divine-human love-making in sensual terms, though it constantly praised Krishna as God. Radha saw Krishna going from one girl to another, 'kissing one and fondling another', appearing to her and then vanishing. She imagined him toying with her friends and remembered their love-making of the past, Krishna 'restless with brimming desire', and 'me who sweated all over my body in the exertion of love'. There was jealousy for God, and dryness and despair which came between early raptures and final union. Eventually Krishna came back to the forest and Radha went to him in her loveliest ornaments. They achieved the heights of bliss in love-play, kisses, and scratches, and Radha took the active part, 'lying over his beautiful body, to triumph over her lover', until 'her thighs grew lifeless' and 'her eyes became heavy and closed'.[28]

[28] G. Keyt, tr. *Gīta Govinda*, 1947.

More sensual than the Song of Songs, yet intended from the first as an allegory of divine-human love in sexual passion, the Gita Govinda is still sung regularly in south Indian temples and its author is considered as a saint. Its relationship to the Kama Sutra is shown by commentaries which explained some of its allusions by reference to the earlier text.

Religious love poetry in other Indian languages also suggested the influence of the Kama Sutra. In the fifteenth century Vidyapati in Maithili and Chandi Das in Bengali described the amours of Krishna and Radha in vivid terms. Vidyapati sang of the swelling breasts of Radha, the first tortures of love, the agony of separation, and the consummation in which the god returned, loosened Radha's dress and 'Krishna makes love the whole night through'.[29] Chandi Das wrote in similar terms of the kisses and passionate embraces of the divine lovers, followed by separation, tears, and reunion.

The religious reformer Chaitanya, in sixteenth-century Bengal, took Krishna as the supreme deity and his highly personal religion was strongly opposed to the pantheism or non-dualism of much Indian philosophy. Devotion to Krishna and Radha was central, expressed in passionate ecstasy and symbolized by erotic love. Chaitanya himself dressed as both Krishna and Radha, in male and female devotions, but some of his followers were said to have had women concubines who were available for the attainment of divine union. To Chaitanya himself only a few couplets are attributed, but his movement helped to develop Bengali literature and is still powerful. Largely owing to Chaitanya sacred sites associated with Krishna were restored, temples were built and millions now visit them on pilgrimage.

Sex and religion inspired much Indian painting, especially miniature paintings of the fourteenth to nineteenth centuries. The Gita Govinda was illustrated in Punjabi and Kangra pictures, depicting in graceful manner the walks, embraces, absence, longing, kisses, and union of the divine pair. Occasionally Krishna and Radha were shown in full sexual

[29] D. Bhattacharya, tr. *Love Songs of Vidyāpati*, 1963.

intercourse, usually with Radha on top, but for the painters
it was the lovers themselves rather than the sexual act which
was the chief concern. In modern times George Keyt, of Shri
Lanka, who translated the Gita Govinda into English, has
painted Krishna as the divine lover, and in pictures of Radha
and the cowgirls he has shown delight in the female form and
in sensual rapture.

More detailed illustrated sexual manuals, often called
Koka Shastras, depicted the coital postures of that popular
text. These were made not only for private interest but often
to demonstrate the virility of the subjects. Indian princes
were shown copulating in the 'act of life', with one or more
women at a time, seated on camels or elephants, shooting
guns, drinking tea, or smoking pipes while copulating. There
were as many varieties as imagination could conceive and
artistry depict, and Europeans, who always seemed to wear
hats and wigs if nothing else, were shown in formal postures
though not so acrobatically as Indians. Religious themes
might become obscured, or transferred to the ruler whose
wives led the cramped life of a harem and whose 'romance'
came through concubines and dancing girls, always young
and beautiful. But religious symbolism appeared in abstract
ways, with the *yoni* represented as a circle, a triangle, or a
lotus in a mandala or pattern. The sexual manuals and erotic
pictures might serve to intensify emotion, revive sagging
powers, prolong intercourse, and perhaps lead to divine love.
Song, music, pictures, and books could bring union with the
deity, though doubtless they often had more immediate and
physical purposes.

Sex and religion also combined in sculpture and architecture. There had long been fertility symbols in the *linga*
and *yoni*, which have been representations of Shiva and his
Shakti down the ages and are seen in countless forms today.
In the medieval period there was a wave of temple-building
which gave India some of its greatest monuments, and many
of the decorations were explicitly sexual. Erotic temple sculpture was ancient and is still practised, notably in Nepal, with
ornamental coital groups. But of the great temples those of
Khajuraho and Konarak are among the most celebrated for
sexual detail. Temples are usually dedicated to Shiva or

Vishnu, and they resemble palaces with the chief deity reigning over a court of gods and dancing girls. The sculptures rise row upon row, the slim-waisted girls helping the movement, while their full breasts and wide hips suggest the bliss of union.

Many positions of sexual intercourse are displayed, and some of the contorted poses seem almost impossible to achieve. In famous friezes the male lover seems to be standing on his head and stimulating three women at the same time. Since the figures all appear to be calm, this may be intended as a ritualistic performance, almost a yoga exercise, to be practised only by accomplished adepts. It has been suggested, however, that the groups may be meant to demonstrate horizontal activities which had to be placed vertically by the needs of architecture. But religious symbolism seems to be obscured sometimes by sculptures with more mundane purposes. The temple of the Sun God at Konarak, in particular, depicts men with erect *lingas* being stimulated orally by female attendants, perhaps to excite their jaded appetites and those of the sculptors' patrons.[30]

These temples shock some modern feelings. When C. G. Jung visited Konarak in 1938 his Brahmin guide told him 'as a great secret' that 'these stones are man's private parts'. Jung was astonished that this obvious fact should be explained, either as if the ignorant European would not have thought of it, or as if the ordinary Indian would be too ashamed to say so. Fortunately Khajuraho and Konarak have been preserved, partly because of the difficulty of access in former times, and although there is still some embarrassment, and exploitation of them for tourists, there is increasing recognition of their importance in Indian culture.[31]

YOGA AND TANTRA

The Sanskrit word Yoga, related to the English 'yoke', came from a root meaning to join together or harness. It came to

[30] A. Watts, *The Temple of Konarak*, 1971.
[31] *Memories, Dreams, Reflections*, 1961, p. 278.

indicate union or conjunction with something, and there were other slight changes according to the context. Yoga included various forms of discipline, of body and mind, practised in order to gain control over the forces of one's own being, to obtain occult powers over nature, and to attain union with God or the universal Being.

Yoga came to be associated especially with the Samkhya, 'enumeration' school of philosophy, providing practical methods for the theory. According to Samkhya there were two orders of reality, Spirit (*purusha*) and Nature (*prakriti*). Spirit was multiple, composed of innumerable monads or male persons, somehow entangled in the material world or female Nature. Salvation would come by complete liberation from Nature and return to the original changeless state, beyond time and space.

Samkhya-Yoga taught a system of yogic practice to gain this liberation by eight stages of discipline, rather like the eightfold path of Buddhism. Some Yoga teachers developed ancient Indian physiological ideas to provide a framework for their practices. There was said to be a great vein of the body running up the spinal column. This vein, called the *sushumna*, contained six 'wheels' (*chakras*) or concentrations of psychic energy at different points along its length. At the top of the vein was a powerful psychic centre, symbolically referred to as a lotus, a female metaphor, situated in the skull and called *sahasrara*. In the lowest wheel was the 'serpent power', a male symbol called *kundalini*, which was generally quiescent. This serpent power should be awakened by yogic exercises and rise up the spinal vein, passing through all the wheels of psychic force to unite with the topmost lotus. By arousing this serpent power the yogi hoped to gain spiritual energy and uniting it with the highest lotus was thought to bring him salvation, though many yogis practised this arousal for the sake of supernatural powers rather than salvation.

The name Tantra was given to the teachings of certain Hindu and Buddhist sects which worshipped divinities or beings especially concerned with sexual energy, and ecstatic cults were inspired by visions of cosmic sexuality. Tantra is not easy to define but the word was applied to scriptures, of which there were many varieties, and it was perhaps derived from a root meaning a 'thread', suggesting the male and female principles of which the universe is woven. The sects were also called Shaktic, from their worship of the female Shakti or divine energy, and also 'left-hand', from their secret or confidential nature and the manner in which the goddess sat on the left of the god. The tantric texts give endless dialogues of these two male and female deities, each alternately teaching and asking questions of the other. These dialogues gave instructions in the meanings and practices of rituals and Yoga.

Tantra emerged in the early centuries of our era as a very popular movement, which affected philosophers and yogis and ordinary people, in rituals, ascetic practices, ethics, imagery, and literature. But it was probably much older; it was non-Vedic and often in opposition to the world-denying philosophy and monistic teachings of Hindu orthodoxy. Instead of suppressing pleasure as a danger or an illusion, Tantra revelled in it, as a means towards the highest goal.

It has been said that Tantra was the rediscovery of the mystery of woman, for every woman became an incarnation of the Shakti, the divine woman and mother. In rituals the female yogi, the yogini, was naked and aroused the feeling of terrifying emotion before the cosmic mystery of creation. Every naked woman incarnated Prakriti, Nature, and in ritual she became the goddess, the Shakti. The Samkhya philosophy was prolonged on to the mythological plane, with the male Purusha, the Spirit, motionless and contemplative, before the active Prakriti, the nourishing power.

For Tantra the greatest energy was sexual and the sexual organs represented cosmic powers, as symbolized in the *linga* of Shiva. Some yogis worshipped their own *linga*, with full ritual, and sexual arousal indicated the coming of the divine presence. The snake was naturally a symbol of sexual power, in the *kundalini* and other concepts. Similarly the female *yoni* was worshipped, and many sculptures depicted not only the female body but its prominent genitals.

Sexual intercourse (*maithuna*) of any kind was treated in a ritual fashion, between husband and wife, or different partners, or with a temple girl. Sexual union was transformed into a ceremonial through which the human couple became a divine pair. The rite was prepared by meditation and ceremonies to make it fruitful, for bodily union alone was not thought to be sufficient to bring salvation. The act of sex was formal and not promiscuous, and coition was not a quick relief in orgasm but a long process in caresses and different postures, for which the Kama Sutra and other manuals were of great help.

Some Tantrics sought to go beyond normal Hindu sexual customs and considered that taboos which were right for others should be broken in order to gain unusual powers. Small groups met, often at night and sometimes in a cremation ground, sitting in a magical circle or pattern. After formal worship they would indulge in the Five M's, forbidden or restricted by others: alcohol (*madya*), meat (*mamsa*), fish (*matsya*), hand-gestures (*mudra*), and intercourse (*maithuna*). In extreme left-hand rituals the woman should be menstruating at the time of intercourse, when her energies were thought to be at the most dangerous peak. Enemies of left-handed Tantra have claimed that sexual orgies were regular practices, and since the ceremonies were secret it was difficult to disprove the charge. Defenders of Tantra said that intercourse only took place between husband and wife, but an eminent anthropologist provided evidence of other pairing and random coupling in recent times. New members were only admitted in pairs, and since the applicant's wife refused to accompany him 'it was not until his adoptive sister' went with him that he was accepted. This pair already broke the taboo of incest, since such coupling was sacrilegious but not casual or promiscuous. At the meeting the women discarded their bodices and put them in a large vessel, and 'during the course of the hymn-singing the presiding Guru called out the men one by one, giving each a bodice chosen at random: this couple then retired behind a curtain to perform the ritual act of sex.'[32] Probably the number of such groups is now very small.

---

[32] G. M. Carstairs in *Aspects of Religion in Indian Society*, ed., L. P. Vidyarthi, 1961, p. 67.

A common feature of Tantric coition was retention of the semen, in *coitus reservatus*. To hasten the ascent of the Kundalini bodily positions were varied with the aim of achieving immobility of breath, thought, and semen. The semen (*bindu*) was thought to have magical power, as in some Upanishadic texts. If it remained in the body, there would be no fear of death, and even if ejaculated into the fire of the *yoni* it could be arrested and returned. Both the man and the woman were thought to be able to recover their vital juices and reabsorb them so as to preserve life. It is possible that this practice came from China, and whether semen could be recovered or not, its retention demanded great discipline and practice. It was also a form of contraception.

Tantric texts stated that all stages of sexual intercourse were to be marked by intonation of texts (*mantras*) uttered many times over the various parts of the body of the beloved. Not merely erotic or promiscuous, this procedure provided ritual and prolonged caressing and union, with the aim of enabling the energies of male and female to pass into each other. The comm-union of separate individuals was transcended in full union and inter-penetration, each holding the other. Sexual union thus provided a complete model of the union of God and the soul. Since it was so profoundly Indian, Tantra not only flourished in theistic Hinduism but also in pantheistic, it developed in ascetic Buddhism, and some of its methods were adopted by even more ascetic Jainism.

REACTION

In recent centuries there have been several reactions against the eroticism of ancient India, particularly its sculptures and *lingas*. The Mughal rulers were Muslims and usually iconoclastic. The fanatical Aurangzeb is said to have destroyed over two hundred temples in one year in one province, and how many were devastated during his fifty years reign is unknown. Not only overtly sexual statues were damaged but simple nudes, such as the stone figures of Jain saints at Gwalior which had their heads and penes knocked off. The heads have now been restored, but in terra cotta.

Muslim puritanism was followed by European. The Abbé Dubois, who wandered about south India from 1792 to 1823 left many vivid descriptions, but stressed what he regarded as the evil side of Hinduism. Of the ubiquitous *linga* of Shiva he wrote of 'obscene symbols', 'the very names of which, among civilized nations, are an insult to decency'. The early British observers sometimes compared Hindu beliefs and sculpture with Greek and tried to understand them sympathetically, but rising enthusiasm for missionary efforts led others to depict everything Hindu in the darkest colours. Thus Charles Grant, in 1792, wrote of 'the immorality, the injustice and the cruelty' of Hinduism, and its adherents 'as depraved as they are blind, and as wretched as they are depraved'.[33]

As late as 1920, the otherwise fair-minded Mrs Stevenson, who tried 'to record the nobler side of ritual Hinduism', spoke of the *linga* of Shiva as 'the livery of his shame', and 'too impure to write here'. But in 1970 the Roman Catholic lay scholar R. C. Zaehner remarked that

Siva's symbol is and always has been the phallus. Once again we are faced with a discordant concord and a concordant discord; but perhaps it is Hinduism that sees the concord more clearly than does Augustinian Christianity, which has been too hasty in its rejection of the 'throes of matter' from which life and consciousness ultimately derive.[34]

The influence of Victorian England was powerful upon educated and reforming Hindus, who adopted the puritanism of their rulers and sometimes became even more rigorous and world-denying. Gandhi's puritan fervour led him to advocate a severe chastity, which deprived his wife of sexual intercourse for years, though the Mahatma is said to have tested his own restraint by sleeping with young girls. A recent Indian writer says that Victorianism, Brahmo

[33] J. A. Dubois, *Hindu Manners, Customs and Ceremonies*, III, V; and see P. J. Marshall, ed., *The British Discovery of Hinduism in the Eighteenth Century*, 1970, p. 42.
[34] S. Stevenson, *Rites of the Twice-born*, p. 374; R. C. Zaehner, *Concordant Discord*, 1970, p. 171.

reformism, and fervent Gandhism 'had overlaid the Indian consciousness with a complex of inhibitions'. Freud, he says, like the Vedas, is hardly ever read by Indian writers, but now the impact of Freud has helped to loosen some taboos, though the tendency is to go to the other extreme. 'The modern gloating over sex is invariably accompanied by a sense of guilt, if it is secret, and by a sense of bravado, if it is brazen. What was a means of legitimate enjoyment to the ancients has become a source of morbid excitement to the moderns.'[35]

[35] K. Kripalani, in *A Cultural History of India*, ed., A. L. Bashani, 1975, p. 417.

Chapter 3

BUDDHIST RENUNCIATION

THE ASCETIC MIDDLE WAY

Gautama, the only Buddha for this present long world era, was married and had a son. He belonged to the warrior-ruler caste, like Mahavira of the Jains, and thus differed in some ways from the priestly teachers. Stories of the Buddha's birth were perhaps current early and written down centuries later, but there are signs of development and growing asceticism in the different accounts.

The Buddha's parents were married, the father being a chief or king named as Suddhodana and his mother the queen Maya. The oldest accounts suggest nothing abnormal about his birth, apart from stressing his noble ancestry. Then legend said that his mother had a dream of a white elephant entering her right side and her womb, and she reported this to the king whose priests interpreted it as indicating the birth of a son, 'a male and not a female', who would become either a universal monarch or a Buddha. The white elephant was the Buddha himself descending from heaven and choosing to enter the queen.[1]

This was not a Virgin Birth, but later stories were more miraculous and anti-sexual. The queen, 'with eyes like a young fawn's' said earnestly to her husband, 'I wish to spend the night away from you'. Then the Buddha-to-be entered her womb, and there arose in her 'no thought of men connected with the senses' for she should not 'be overcome by any man of passionate heart'. When the child was born she

---

[1] Digha Nikaya, 2.52, and Nidanakatha.

41

did not give birth sitting or lying down, as other women do, but standing, which, as a French scholar remarks, may have been more elegant but was probably no more comfortable. The baby was said to be 'born clean, unstained with liquid, unstained with phlegm, unstained with blood, unstained with any filth', and streams of hot and cold water fell from the sky to wash the mother and child.[2]

The narratives of the birth of the Buddha have been compared with the stories of the Virgin Birth of Christ, but the two traditions are entirely different. There is no trace in the Gospel tales of the anti-material and anti-sex feeling that develops in the Buddhist stories. This feeling appeared in a further element in the narratives which said that the mother of this Buddha, and mothers of all Buddhas, died seven nights after giving birth, so as to avoid further sexual intercourse with their husbands, 'because it is not fitting that she who bears a Peerless One should afterwards indulge in love'. Despite this anti-sexual prejudice, the monkish writers of the stories could not resist giving sensual details, for example about the queen lying on her bed, 'radiant, alluring, and gleaming as with the sheen of gold . . . O woman, whose belly, with its bright streak of downy hair, curves like the palm of the hand.'[3]

The newborn Buddha bore the thirty-two marks of a Great Man, which included wheels on the soles of his feet, projecting heels, long toes and fingers, tender hands and feet, long arms, soft skin, rounded bust, forty regular teeth, intense blue eyes, mole between eyebrows, head like a royal turban. His male organs were enclosed in a sheath, and to a doubting inquirer one day the Buddha 'arranged matters by his wondrous gift that the Brahmin saw how that part of the Blessed One that ought to be hidden by clothes was enclosed in a sheath.'[4]

The Buddha-to-be married the beautiful Yasodhara who bore him a child named Rahula, who later became a monk and followed his father. Legends make much of the luxury of Gautama's life, in a palace, with many male and female

---

[2] Mahavastu, 2.1 f., and Lalita-vistara.   [3] Mahavastu, 2.1.
[4] Digha Nikaya, 3.106, etc.

42

attendants. But all agree that at the age of twenty-nine Gautama was shaken out of his ease at the unexpected sight of old age, disease, death, and an ascetic. He abandoned home and family by night, sickened, according to later story, by the abandoned postures of sleeping male and female servants. He became a wandering ascetic, trying Yogic and other teachers, fasting till he became like a skeleton, and then taking food. He decided to follow the Middle Way, between the extremes of sensuality and asceticism, and this is formally the Buddhist position, rejecting the self-torture of extreme Indian ascetics. The Buddha was tempted by both Kama and Mara, love and death, but resisted them and took up his world-renouncing teaching. He never returned to his wife and, after his enlightenment, other men joined him to form a band of monks or brothers.

World-denial, in various forms, had long been practised in India, not only by those who reached the last stages of life, but by men who were more intent on knowledge than on love. Early Upanishads said that those who know the Soul overcome 'desire for sons, desire for wealth, desire for worlds, and live the life of mendicants'. They became disgusted with both learning and childhood simplicity and became ascetics.[5] Buddhism developed the celibate life, but with emphasis upon community. There was the solitary Buddhist, 'like a lonely rhinoceros', but the brothers lived in community and this Sangha, the Order, became one of the three elements of Buddhism, after the Buddha and the Dharma or doctrine.

There was criticism of this world-renouncing life, which the Buddha countered in an early dialogue, pointing to the troubles and temptations of family life:

> The household life is full of hindrances, a path for the dust of passion. How difficult it is for the man who dwells at home to live the higher life in all its fullness, purity, and perfection. Free as the air is the life of him who has renounced all worldly things.[6]

Therefore the monk, even if he had been a slave before,

[5] Brihad-aranyaka Upanishad, 3.5.   [6] Digha Nikaya, 1.62 ff.

would be honoured by his former master, who would provide him with a lodging place, medicine, robes, and a bowl, and watch and guard him. The monk has put aside all temptations, restrains evil inclinations, masters his faculties, and looks at the inner meaning of all his actions. He is free from the deadly taint of lusts, ignorance, and pain. The basic Four Noble Truths declared that birth, life, and death were painful; the pain came from craving, combined with pleasure and lust, finding pleasure here and there, the craving for passion, the craving for existence, the craving for non-existence. Pain would cease by overcoming this craving, which resulted from following the Noble Eightfold Path of moral and spiritual discipline.

Buddhism was primarily for monks, and it was said that the monk was the only true Buddhist. There were lay followers from an early date, men and women, but there was no doubt of monkish and male superiority, and lay people who did not join the order in this life could hope to be reborn as monks in the next, though the trials of the monastic life soon appeared.

CELIBATE AND SEXUAL POWER

Like King Brihad-ratha in his world-disgust, in the Maitri Upanishad, the Buddhist monk was taught to contemplate the body 'encased in skin and full of various impurities: nails, skin, teeth . . . stomach, excrement, bile, phlegm, pus, blood, sweat, tears.' He should reflect on it like a cattle-butcher who, having slaughtered a cow, might sit displaying its carcase at the crossroads; or he might contemplate a body in various stages of decomposition in a cemetery. The aim was to bring detachment, beyond pleasure or pain of the body and its cravings.[7]

The danger of women to monks was illustrated by some well-known advice said to have been given by the Buddha himself. His disciple Ananda asked: 'How are we to behave towards women?' The Buddha answered: 'Do not look at them.' Ananda objected: 'But if we have to see them, what

[7] Majjhima Nikaya, 1.55 ff.

shall we do?' The Buddha said: 'Do not talk to them.' Ananda persisted: 'But if they speak to us, what shall we do then?' The Buddha warned: 'Keep wide awake, Ananda.' [8]

There were several reasons for the danger of women to monks. Sexual relations would bring attachment that would distract the monk not only from his vow of chastity, but from the search for liberation. Moreover, children might be born and family life would bring further ties. Monks therefore denounced sexual intercourse as 'bestial', and looked on women with fear and contempt. The monkish life was called Brahma-charya, literally the conduct of a Brahmin or holy man, but it was regarded as synonymous with chastity or abstention from all sexual intercourse.

Sex was feared because it could be a rival to that calm and joy which the monk sought by his way of self-denial. A lover would find peace and fulfilment in the raptures of coition, which countered those of nirvana. Sexual intercourse and self-denial have common elements, single-mindeness in method and losing self in unspeakable bliss, but the monks believed that theirs was the highest way and that it excluded any concessions to lower passions. For about a thousand years any variation of monkish celibacy was excluded, and the rule still holds in most places.

Elite Buddhism was a discipline that required celibacy, which was perhaps necessary in community living. In Hinduism priests were married but early Buddhism had no priests, said little about marriage and nothing positive on sexual life. Many detailed regulations restricted a monk's behaviour towards the women he met, or the nuns that he taught, and breaches of chastity meant expulsion from the monastic order.

But sex could not be banished so easily, and since monks and nuns had sexual instincts it was understandable that these made themselves felt, even when disguised or reprobated. The first long division of the canonical scriptures of the Theravada Buddhists was the Vinaya Pitaka, the 'discipline basket', which gave the rules governing the Sangha. Every fortnight 227 disciplinary rules were recited, and a

[8] Digha Nikaya, 2.141.

45

number of them dealt with sexual temptations or offences. It was wrong to masturbate, or to persuade a woman to sexual intercourse. A monk should not touch a woman, hold her hand or arm, stroke her hair, caress or rub any part of her upper or lower body. He should not speak to a woman of the gift of her body as a supreme service to offer intercourse to monks. He should not sit alone with a woman in the open, or even act as a go-between for women and men, for meetings, adultery, or marriage.

The detail of these rules indicated the nature and strength of temptations, and stories used to illustrate the rules went into prurient particulars. The English translation of the Vinaya is in five volumes, and the spinster lady who translated it put asterisks for verses that were too blatant, but many verses were left to reveal both the temptations of monks and the fascination that imagination of adventures had for them. Many discussions consider the behaviour of monks and nuns, their natural functions, accidental encounters, and errors that could come from the most unlikely possibilities. Thus it was permissible to emit semen in a dream or unconscious state, but not when awake. To illustrate this a monk was said to have been asleep by the wayside and was noticed having an erection by passing women. Each of them lay with him in turn but he did not wake up, and they said 'he is a bull among men'.[9] Such stories may border on pornography, but the Vinaya is too long and obscure to be taken up by modern printers of erotica. Differently from Hinduism these erotic verses do not serve any theological purpose of illustrating human or divine union through sex.

The Jataka tales, 'birth stories' of the Buddha, suggest sexual temptations even for the founder of the religion. When in a former birth he was an ascetic, he went to Benares on a begging-round and was taken to the king's palace. There he lived for sixteen years and when the king went away his queen ministered to the ascetic's needs. One day she prepared his food and having bathed lay down to rest, but when the monk came in the queen started up and her dress

[9] Vinaya, 1.10,21.

46

slipped off so that he gazed on her naked beauty for pleasure's sake. Lust kindled within him, insight deserted him, and he was like a tree felled by an axe. When the king came back the ascetic confessed his desire for the queen, and the king gave her to him. But he privately charged the queen to save the holy man, which she promised to do. She demanded a house and the monk found a tumble-down hut. She told him to clean out all the filth, get a bed, a stool, a rug, and a thousand other things. Finally as he approached her on the bed she seized him by the whiskers and said, 'Have you forgotten you are a Brahmin and a holy man?' Then the hermit came to himself, returned the queen to her husband, and flew through the air to the Himalayas, there to meditate in unbroken insight. 'I was that hermit', the Buddha is reported to have said.[10]

Many other stories were told of the fascination and repression of sexual desires in Buddhist monks, but the constant orthodox theme was that sexual indulgence was both against the Buddhist law and undermined spiritual power. In the Vinaya it was said that it would be better for a monk's *linga* to enter a snake or a burning fire (both sexual symbols) than to enter a woman's *yoni*. Taboos were extended beyond women, old and young, mothers and children, to prohibition of handling female animals, including sleeping under the same roof with them. When teaching a mixed congregation a monk should keep his fan before his eyes so as not to be tempted by the sight of an attractive woman. Even a monk's own mother should not be touched; if she fell into a ditch he must offer her his stick but not his hand, and think that he was pulling out a log of wood. Taboos, as so often, might either kill natural affection or kindle suppressed desire.

Especially in southern Asia Buddhist monks have kept the rules of celibacy to this day, in those lands where monasteries are still allowed to exist. In Burma, says an anthropologist, few cases of unchastity can be discovered and since monasteries are open to the public it would be difficult to hide derelictions. In one instance a monk had an affair with a woman who became pregnant and had an abortion. Five

[10] Jataka, 66.

years later the monk was still carrying out penitential ceremonies. But in general Burmese monks seem to respect
sexual prohibitions, both homosexual and heterosexual.
With the open monasteries, those who are sexually frustrated
can leave the order, so that those who remain would say that
they are not bothered by the vows of chastity.[11]

European writers in the last century made similar remarks
about the fidelity of Burmese monks to their vows of sexual
abstinence and declared that breaches of them were rare
occurrences. But some writers declared that in neighbouring
Siam (Thailand) women went to the monasteries and sometimes even lived there, while in Ceylon (Shri Lanka) homosexuality was said to be not infrequent between monks and
novices and laymen, and one monk kept a mistress in the
monastery grounds. On the other hand, a modern anthropologist in central Thailand had the impression that most
monks there kept closely to the rule of celibacy, rare in-
fractions were reported, and the few who were accused of
showing too much interest in 'the way of the world' had
simply been seen to give occasional bold glances at young
women in the congregation. Thai monks said it was the
Burmese monks who took girls to football matches and
places of entertainment. The remarks reveal national
prejudice or envy.[12]

In some northern Buddhist countries the opinion slowly
grew that sexual life was not incompatible with that of a
monk, especially outside community monasteries. From
about A.D. 500 there were married monks in Kashmir, and
with the rise of Tantra the numbers of married monks increased there and elsewhere in northern Buddhism. The
founder of Buddhism in Tibet in the eighth century, Padma
Sambhava the 'Lotus-born', who was regarded as a second
Buddha, received a wife from the Tibetan king and had at
least two principal wives whose names have been recorded.
In the eleventh century the Buddhist translator Marpa in
Tibet was married and had eight other female disciples who
were called his spiritual consorts. In China and Japan

---

[11] M. E. Spiro, *Buddhism and Society*, 1971, pp. 366 f.
[12] J. Bunnag, *Buddhist Monk, Buddhist Layman*, 1973, pp. 30 f.

Buddhist priests who officiated at temples were often married, but monks in communities generally remained single.

The formation of orders of female ascetics, or nuns, was debated from the early centuries, and several legends tell how the admission of women to the Buddhist Order came to be allowed. The Buddha's widowed aunt wished to renounce the world, but he refused her three times. Then at the request of his chief disciple Ananda the aunt was admitted, but with the warning that if women had not received this discipline the Order and Doctrine would have lasted a thousand years, but as women had come in it would only last five hundred years. But this was too pessimistic, since Buddhism has now lasted 2,500 years and women probably make up most of its adherents, as they do in other religions.

BUDDHIST TANTRA

Tantra, we have seen, was ancient in India and although the oldest complete texts that survive are Buddhist, dating from about the fifth century A.D., the practices went back to pre-history with revival and systematization later. Buddhist Tantra is generally known under the name of Vajrayana, 'Diamond Vehicle', which appeared at the beginning of the fourth century and reached its full flower in the eighth, though tradition says that it was first introduced earlier than these dates. It is noteworthy that Buddhist Tantra developed in north and eastern India, and other places which were not fully Hinduized, and in Tibet and south-east Asia.

The Diamond Vehicle was a new revelation of the Buddha's doctrine, so it was claimed, adapting the doctrine to the needs of the expanding religion. It was said that a king went to the Buddha and asked him for a Yoga that could save men in this present dark age. The Buddha revealed to him that the universe is contained in man's body, with its central sexual powers, and taught him how to attain liberation by proper methods. Some Tantric schools rejected basic Buddhist practices such as meditation, since they sought simplicity and return to nature. One text that aimed at the achievement of Buddahood said that sensuality was permitted, eating flesh, lying, stealing, and committing adultery.

Most Buddhist Tantras could be divided into four interests, the first two kinds of text being concerned with rituals and the others with yogic ways of attaining supreme truth. But in fact most Tantric texts include ritual matter as well as instruction in Yoga and philosophical speculation. Different kinds of Tantra are meant to be suited to activities and temperaments, monkish and lay.

Early Buddhism was male-dominated, but with the incorporation of Tantra at both popular and philosophical levels the feminine element became very important. Tara, a saviour goddess who helps to cross the river of death, was a popular Buddhist creation that had no clear Hindu origin. Prajnaparamita, the 'perfection of wisdom', was a goddess of philosophical interest, a female principle placed alongside the Buddha or called 'the Mother of all the Buddhas'. As a mother her full breasts were emphasized, but she was also called an 'untouched' virgin.

Buddhists, like Hindus, distinguished 'right-hand' from 'left-hand' Tantra. Hindu right-handers devoted their attention to the male principle and left-handers to the female. In Buddhism it was the left-hand form that was particularly concerned with sex. Right-hand Buddhist philosophers taught apparently contradictory or meaningless doctrines, such as 'transmigration is nirvana', or 'there are no beings', or 'reality is emptiness'. Left-handers declared that 'the passions are the same as nirvana' and they should not be suppressed. Tantrists adopted magic spells (*mantras*), ritual gestures (*mudras*), and dances which were 'singing with the body'. Magical circles (*mandalas*) were used as aids to concentration, and great numbers of terrifying deities and demons were added to the pantheon of gracious Buddhas and gods.

Left-handed Tantra said that adepts should 'cultivate all sensual pleasures' in order to become Buddhas. Taboos should be broken, not only by eating the flesh of forbidden animals, but mingling it with ordure and urine, to stimulate the senses by repugnance. In the sexual mysticism of Vajrayana, the 'thunderbolt' is the phallus, and in other texts it is the Buddha himself as phallus, while Sukhavati or paradise is the *yoni*, and five vital airs represent semen.

Left-handed Tantra worshipped female and male deities in the embraces of union, and the Buddha himself was said to be united in continuous sexual activity, for he revealed the truth that 'Buddhahood abides in the *yoni*'. Buddhist art, like Hindu, often expressed this divine union in sexual imagery. Buddhas and Bodhisattvas, 'beings of enlightenment', were painted and sculpted in the act of coition. Innumerable forms of the Buddha were embraced by their Shaktis to provide examples of intercourse, and in Tibet they were the *yab-yam*, 'father-mother', the progenitors. Nepalese and Tibetan brass images represented such sexual deities, and in initiations into some Tantric sects the image of the goddess with open legs was placed in the lap of the male initiate to indicate symbolic intercourse, and power was believed to be transmitted from the image which was filled with divine energy. The well-known painted banners from Tibet often display a central half-nude female figure, surrounded by other beings or pleasures, who is the goal of Tantric exercises.

Tantric adepts were supposed to attain salvation by copying this divine pairing with female partners, and there were antinomian relationships with another's wife, a virgin, sister, daughter, mother, or even supposed demonesses. In ancient Indian Buddhist Tantra, which survived later in parts of Bengal and Tibet, there were special women who initiated men into Tantric mysteries. Called *dakinis*, 'sky-walkers', or Yoginis, female yogis, they were often depicted in horrific forms as naked females like Kali, or devils. Such women, filled with supernatural power by ritual and intercourse with initiated Tantric men, were thought to pass on their powers to men seeking initiation by intercourse. Padma Sambhava is said to have received his initiation from intercourse with a Dakini in a cremation ground, a place that had a morbid association of power as in Hindu Tantra.

Tibetan yogic books gave exercises whereby the devotee might obtain communion with the divine being. He had to imagine himself to be the very Vajra-Yogini who was red in colour, naked as being divested of all things in the world, in the full bloom of virginity as unsullied, dancing with her right leg uplifted and left foot treading down a male body which was all things of ignorance. Vajra-Yogini had three eyes, the third being the eye of insight; her right hand uplifted a shining curved knife to cut off all disturbing thoughts, and her left hand held against her breasts a human skull full of blood to symbolize renunciation of the world. She had a tiara of human skulls and a necklace of fifty human heads dripping blood, said to demonstrate the severance of *samsara*, the round of births and deaths. The horrific aspects of this red or black naked goddess, like the Hindu Kali, was meant to inspire fear and renunciation, and to break all taboos, and it probably went back to prehistoric India and Tibetan beliefs and rituals.

Meditation on the Yogini led to the arousal of internal psychic heat. The letter HAM, the Tibetan personal pronoun 'I', was to be visualized as white in colour like the semen which was thus set in activity. The goddess Kundalini, the serpent power, rose from her slumber below the navel to union with her lord in the highest centre of the skull. The symbolism was of phallic erection but retention of the semen or 'moon-fluid', which was transmuted into psycho-physical power. The aim, as in other sexual exercises of similar nature, was not ejaculation but transmutation of physical energy into spiritual, and thereby attainment of salvation.[13]

Vajrayana seems to have become extinct in India by the twelfth century, as Buddhism declined before reviving Hinduism. But its teachings had been taken to Tibet where they merged with indigenous cults, and from Tibet they spread to Mongolia and to China where Kubla Khan and the Mongol rulers who succeeded him in China were devotees of Tibetan forms of Buddhism. Kubla Khan surrounded himself with Tantric experts who invested him as World Monarch according to Tantric ceremonies. Chinese Confucian scholars gave horrifying accounts of sexual orgies, and even bloody sacrifices of women among the Mongols, but they were certainly prejudiced against them and probably ignorant of their beliefs and rituals. The relation of Bud-

---

[13] W. Y. Evans-Wentz, *Tibetan Yoga and Secret Doctrines*, 1958, pp. 173 f., 191 f.

dhism and Tantra to Chinese Taoism and Japanese Shinto will be discussed later.

Buddhist Tantra infiltrated south-east Asia and was found in Java about the tenth century. Sexual rites were known to have flourished in some of the numerous monasteries of Pagan in Burma in the eighth century, and in Cambodia Chinese travellers in the fourteenth century claimed that every maiden had to be deflowered by a Buddhist priest before her marriage, though whether this was by Tantric ritual or a more ancient custom was not clear. Most of this has gone, and what remained in Tibet has been swept away by Chinese Communism. Orthodox Buddhists were shocked by the deliberate breaking of taboos by Tantrists, and their horrific imagery. At its best Tantra helped to sacralize sexual intercourse and treat it as a copy of cosmic unity.

LAY MARRIAGE AND MORALITY

Buddhism began with disciples gathered round the founder to form the nucleus of the monastic Order, yet the movement rapidly developed into a religion in which a large following of men and women supported the monks. Many Buddhists, lay and monkish, knew little or nothing of Tantric practices even in their heyday, and the detail of Tantrism must not be allowed to create a picture of morbid fascination with the erotic or horrific.

In the Vedic epoch marriages were arranged by the father, and although under Buddhism women gained some degree of independence, in having more voice in choosing a partner, yet the texts still spoke of fathers 'giving their daughter in marriage', and boys also were said to marry with the parents' consent. A princess Kanha asked her mother to persuade the father to hold a 'self-choice', an assembly 'to choose me a husband'. If the father failed to find a suitor, after the girl had waited his instructions for three years, she might 'choose for herself a husband of equal rank in the fourth year'.[14]

Daughters had often been regarded as unwelcome burdens, although the Buddha comforted a king whose wife had

[14] See I. B. Horner, *Women under Primitive Buddhism*, 1930, pp. 19 ff.

given birth to a girl by saying that she might 'prove an even better offspring' than a boy, but this was an isolated saying and the old view seems to have prevailed. Yet female infanticide, practised in some other parts of the world, never seems to have taken hold in India and certainly not in Buddhism. There is little reference to child-marriage in the texts, though one story does tell of a girl of less than twelve years of age who was a bride when she was ordained into a nunnery. It was said that when girls reached the age of sixteen 'they burned and longed for men', and parents hastened to marry them off before they eloped as occasionally they did. Women who were immured in the house might have intercourse with a slave, as one was recorded to have done when 'she was maddened with the madness of youth and lusting for a man'.

The arrangement of a marriage was a family affair and Hindu customs of consulting astrologers were firmly rejected by Buddhists, at least in classical times. Fathers provided the dowry, chiefly of clothes and jewelry, and the rich added many other items including slaves and cattle. Bride-prices were also paid to the parents and in a second marriage the amount was reduced. The wedding ceremony was domestic or civil, without the ritual of a priest. The wedding took place at the bride's house and lasted several days according to the wealth of the parents. It was composed of joyousness and feasting and afterwards the bride went with her husband to live in his parents' house, where she was supposed to be 'pleasant' and show deference and humility to her parents-in-law.

In modern times marriage customs vary from country to country, and even in regions. There is often a blessing performed by monks, followed by a civil ritual and in some countries registration is required by law. This may obtain more in towns than in villages, and in the latter 'the fact of the young couple's living together being the seal of marriage in the eyes of the community.'[15] Legally the husband is head of the union, but either partner may possess private property as his or her sole concern.

[15] J. de Young, quoted in J. Bunnag, *Buddhist Monk, Buddhist Layman*, p. 15.

The mutual duties of husband and wife are sketched in one of the main bodies of Buddhist scripture. A husband should minister to his wife in five ways: respect, courtesy, fidelity, giving her adornments, and allowing her authority in the household. A wife should minister to and love her husband in five ways: by doing her duties well, by being hospitable to relatives of them both, by fidelity, watching over his good, and industry in all her business.[16]

The customs of developing Buddhism were illustrated by the Jataka tales. Polygamy must have been found chiefly among the rich and powerful, but it was approved. There was a Brahmin who had four daughters and four suitors came who were respectively handsome, old, noble, and virtuous. To which suitor should he give his daughters? He asked a Buddhist teacher who gave the standard answer that virtue was the supreme quality, and the man gave all four daughters to the virtuous suitor. The tale ended with the Buddha saying, 'that famous teacher was I myself'.[17]

Although Buddhism cut across caste divisions and gave more freedom to women, yet traditional male dominance remained. Woman was the inferior and possession of man and, as one authority said, she was 'never fit for independence'. The Vinaya gave a list of ten kinds of wives: those who were bought for money, those living together voluntarily, those to be enjoyed or used occasionally, those who had given cloth, those who provided the house with water, those with a head-cushion to carry vessels, those who were slaves and wives, those who were artisans and wives, those who were prisoners of war, and those who were temporary or momentary wives. The trials of polygamy were mentioned elsewhere in verses which speak of a woman's lot as woeful 'when sharing homes with other wives'.[18]

Chastity of both husband and wife was taught in classical Buddhism. The Buddha is said to have commanded: 'Let the wise man avoid an unchaste life, as he would a burning heap of coals; if he cannot live a life of chastity he should not

[16] Digha Nikaya, 3.190.
[17] Jataka, 200.
[18] Vinaya, 3.139 f. See I. B. Horner, *Women under Primitive Buddhism*, pp. 43 ff.

transgress with another man's wife.' And again, 'As rain breaks through an ill-thatched house, so passion makes its way into an unreflecting mind.' For, 'Passions are of small enjoyment and productive of pain in a wise man.' But such exhortations are as much or more for monks as for laymen.[19]

One Jataka tells of a princess who was very virtuous, but one day looking out of a window she saw a handsome viceroy sitting behind the king on an elephant's back. She fell in love with him and thought that if her husband were to die she could marry the viceroy, but she remembered that she was married and was full of remorse. She told some Brahmins who assured her that it was a small thing, and she had not broken virtue by the mere thought. However, she herself continued to think that her virtue was no longer perfect.[20]

All Buddhists should keep the Five Precepts (*silas*): not to take life, steal, indulge in immorality, lie, or become intoxicated. The same Jataka said that all the inhabitants of the kingdom kept the Five Precepts, including strangely enough the courtesans. For the interest of the story a virtuous courtesan was approached by Indra, king of the gods, in the form of a youth. He gave her a thousand pieces and promised to return soon, but he went back to heaven for three years. The courtesan remained faithful to him 'for honour's sake', spent the money and gradually became very poor. Finally she went to the chief justice, told her tale, and he said that she must return to her former employment of prostitution. She went away and at once a man came to offer her another thousand pieces, and as she held out her hand Indra revealed himself and praised her fidelity. The courtesan complained that her virtue was not perfect, since she had held out her hand again, but the Brahmins assured her that merely to hold out the hand was not a breach of virtue.

The Buddha himself had some women followers and one of them, Ambapali, was a courtesan. She invited the Buddha to a meal and 'he gave his consent by silence'. He arrived with many followers, who were all fed, and the courtesan brought a low stool to sit beside the Buddha, who instructed and gladdened her with religious discourse and then

---

[19] Dhammapada, 13 f., 186 f.   [20] Jataka, 276.

departed. The Buddha told his disciples that the enlightened mind was set free from the Intoxications of Sensuality, Becoming, Delusion, and Ignorance. [21]

Other courtesans appeared in Buddhist story, sometimes living in groups and at times very wealthy. Their occupation was never openly condemned, it was listed at the end of names of other professions and was regarded as low rather than blameworthy. In Buddhist and general Indian thought a prostitute was working out her *karma*, like all other ranks in society. She must have been reborn to that low condition by some offence in a previous life, but she need not remain there, and by virtuous practice she could rise to a higher state in the next life. A few courtesans became nuns, but of only one is it recorded that she was converted by the preaching of a leading monk who spoke of the foulness of the body, to deter her from continuing in her profession.

The Order of nuns provided scope and refuge for varieties of women, perhaps especially those who were 'not guarded', such as orphans, unmarried, widows, and virtual widows. But in the early years, at least, numbers of nuns came from royal, wealthy, or learned families. Some became renowned preachers, and of one it was said that she was 'sage, accomplished, shrewd, widely learned, a brilliant talker, and of goodly ready wit'. Collections of verses or psalms illustrate these feminine qualities, but the male monks were always considered to be the more important of the two Sanghas. [22]

Western studies of Buddhism in the nineteenth and twentieth centuries often represented it as an ethic rather than a religion, with no god or soul, and with a simple humanistic morality for intellectuals. Such an exposition was unable to explain how Buddhism had become the religion of half Asia. Temples were written off as superstitious joss-houses, and Tantra was unspeakable. Only in recent years has the wide range of Buddhist religion, and its attitudes on sex, been recognized.

Popular Buddhism reflected many general Indian attitudes in its art. The Buddha was the central figure in groups of sculpture, though in the first centuries he was not represented

---

[21] Digha Nikaya, 2.97.　　　　[22] I. B. Horner, op. cit., part II.

in human form but symbolized by his shoes or umbrella. But the attendants on the Buddha resembled those of many Hindu sculptures, kings and commoners, men and women, animals and demons. The great stone gateways at Sanchi still display a wide range of human figures, including naked girls with clearly marked *yonis*, and in more monastic centres at Ajanta and Amaravati there are also many lush female figures. Gods constantly appear, to the confusion of those who imagine Buddhism to be atheistic, and to this day Vishnu, Shiva, or Indra may be seen in many Buddhist temples in southern Asia, though the Buddha is in the centre.

In its art and mythology Buddhism encompassed a great variety of human experiences, and if this was so of the Theravada Buddhism of India and southern Asia, even more complex pictures were provided by Buddhist life in China, Japan, and northern lands where Buddhism long flourished, and Buddhist attitudes to sex will be considered further in these places.

Chapter 4

OTHER INDIAN TRADITIONS

Hinduism was not a closely organized religion, with a distinctive church, but a vast complex of religious currents. Many of its attitudes affected other Indian religions, and some of its models and precepts were generally accepted. Rama and Sita were patterns of courage and marital virtue far beyond India, and Hindu gods appeared still in reforming Indian traditions.

JAIN ASCETICISM

The Jain religion appeared in history as parallel to Indian Buddhism, though the Jains teach that their religion is eternal. Mahavira, the 'Great Hero', who was roughly contemporary with the Buddha, was said to be the last in a long line of teachers going back millions of years. These were the twenty-four Jinas, 'conquerors', or Tirthankaras, 'fordmakers'. Some scholars think that both Jainism and Buddhism preserved elements of ancient Indian religion, perhaps partly from the Indus valley, differently from the Vedic Aryan religion.

Jainism was more ascetic than the Buddhist Middle Way, and has been called atheistic or perhaps trans-theistic, since it included Hindu gods but regarded them as subservient to the Jinas. Mahavira's parents were followers of the previous Jina, Parshva, and having lived very strict lives they finally rejected all food and starved to death, so avoiding creating any more *karma* which might have brought them back to rebirth. They became gods, and should eventually reach absolute perfection, the end of all misery, and Nirvana.[1]

Mahavira is said to have descended from heaven, like the Buddha, and to have taken the form of an embryo in his mother's womb. Then the monkish writers of this ascetic story could not resist embellishing it with sensual details. The mother had fortunate dreams in which, among fourteen auspicious signs, she saw the great goddess Shri whose body was described in glowing terms in the scriptures. She had hands and feet like lotus leaves, round legs, dimpled knees, fleshy thighs each like an elephant's trunk, large and beautiful belly with a lovely row of soft black hairs, waist which could be encompassed with one hand, cup-like pair of breasts, lovely face and eyes pure as water lilies, and glossy black braided hair.[2]

After such a panegyric it was not perhaps surprising that the mother-to-be went from her bed, 'with a quick and even gait like that of the royal swan', and told her husband of the dreams. But there had been a curious preliminary wherein the child had been conceived first by Brahminical parents, enjoying 'the noble permitted pleasures of human nature', but the embryo was then transferred by the gods to the womb of Kshatriya rulers, taking off 'all unclean particles' and bringing forth 'the clean particles'. Such baby-swapping occurs also in the stories of Krishna, but here it seems designed to give the child the best of both Brahminical and Kshatriya worlds.

The Kshatriya husband, Mahavira's eventual father, called the interpreters of dreams to explain his wife's experiences and they declared that, like the Buddha, the child would become either a universal monarch or a Jina. The boy was born into this warrior-ruler caste, perfect and beautiful, on a night when 'there was a divine lustre originated by many descending and ascending gods and goddesses', in the universe resplendent with a single light. He grew up clever, of great beauty, controlling his senses, lucky and modest. He lived the life of a householder, married a wife named Yashoda, who bore him a daughter. But the Digambara sect deny this marriage, since celibacy is a higher state.

[1] Acharanga Sutra, 2.16.   [2] Kalpa Sutra, 36.

At the age of thirty, after his parents had gone to the world of the gods, Mahavira renounced his kingdom and possessions to the care of his brother. He went to a public park, putting off all his ornaments and finery, and plucked out his hair in five handfuls, to enter the state of homelessness. Mahavira wore clothes for a year and after that he walked about naked and accepted alms in the hollow of his hand. He neglected his body for more than twelve years and then attained enlightenment, becoming a Jina and knowing all things of gods and men. He wandered about, teaching the law to gods and men, till he died at the age of seventy-two and became perfect, a Buddha, finally liberated and freed from all pains in the ultimate nirvana.

Jainism was an ascetic order, and though it required a large body of lay supporters only a monk could attain full salvation. To gain nirvana a man must abandon all ties, including his clothes. The major schism in Jainism came over the rule of nudity, when those in the colder north of India allowed the wearing of white clothes and were called Shvetambaras, 'white-clad', whereas the more extreme always went naked and were called Digambaras, 'space-clad'. Today, however, most monks of both sects wear robes, at least in public. A contemporary religious movement, the Ajivikas, forbade the wearing even of a loincloth. But Jains still say that nudity is essential to reaching nirvana, which few if any attain nowadays.

Jain monks and nuns live extremely ascetic lives. Their hair is not shaved off but pulled out at initiation, they have frugal meals, with many fasts, and numerous monks have starved to death. Monks take five vows: not to take life, steal, kill, indulge in sexual intercourse, and possess property. Not killing (a-himsa), now popularly interpreted as non-violence, is taken more literally than in other Indian religions. Both eating meat and taking any life is forbidden to monks and laity. Jains could not become farmers, for fear of killing animals and insects, and so many were traders and became rich. Monks were meticulous in avoiding taking life, wearing masks over their mouths and brushing their path to save insect life.

Jain monks were warned severely about the dangers of

61

women: 'those who have intercourse with women are no better than householders, but they are no monks.' Since 'in this world men have a natural liking for women' one who renounces them will easily perform his monkish duty. For a wise man knows that 'women are a slough' and 'like a poisoned thorn'. Temptations were detailed: women make up to monks with clever pretences, they show off their armpits or lower parts of their bodies, or tempt them to comfortable beds. A monk should not look at them or walk with them, go into houses without a companion, talk to his own daughters or slave-girls or would-be female disciples alone. Those who talked to women had already ceased to meditate, though some tried to excuse themselves by taking a middling position, between a householder and a monk. Severe penalties, with amputation of hands and feet, were promised to adulterers though these probably referred to civil punishments rather than to monastic.

The bondage to which a monk could be reduced if he was absorbed with passion for a woman was vividly described. When she had captured him she would send him on all sorts of errands: fetch some fruit, get wood for the fire, paint my feet, rub my back, buy me some perfume, ornaments, powder, umbrella, slippers, fan, comb, ribbon, looking-glass, tooth-brush, betel-nut, needle, thread, chamber-pot, basket, mortar, water-pot. The man would be told to dig a privy, hold the baby, get up in the night to lull it to sleep like a nurse, wash clothes like washermen, and become a beast of burden or a mere nobody. A monk should therefore pay no attention to the entreaties of women, knowing that their dangerous pleasures will not be for his benefit, so he will abstain from women, and from any unnatural crime, vanquish sin and delusion, and wander about till he reaches final liberation.[3]

Women are therefore 'female demons', on whose breasts grow two lumps of flesh. They continually change their minds, entice men, and make a sport of them as slaves. Putting an end to such delusion was illustrated by the story of king Nami who having enjoyed, 'in the company of the

[3] Sutra Kritanga, 1, 4.

beautiful ladies of his seraglio', excellent pleasures to match those of the heavens, became enlightened as to their true nature. He placed his son on the throne, retired from the world, and resorted to a lonely place. The kingdom was in an uproar and Indra, king of the gods, came down to see what was happening. He told Nami that his palace was on fire and he should look after his harem, but Nami calmly replied, 'We are happy who call nothing our own, when the city is on fire nothing is burned that belongs to me; to a monk who has left his sons and wives, and who has ceased to act, nothing pleasant can occur, nor anything unpleasant.' Indra praised the royal ascetic, revered him again and again, and flew back to heaven 'with his crown and earrings prettily trembling'. [4]

Jain laymen took five vows similar to those of monks. The vow of chastity was interpreted to mean that a man should be faithful to his wife at all times, not consummate marriage with a young child, or form a temporary connection with a woman whom it was impossible to marry. Brahmin priests were domestic chaplains to Jain laymen and officiated at their wedding, birth, and death ceremonies. The age of marriage varied among Jain sects; generally daughters were married at about fourteen or fifteen, to boys of nineteen or twenty, though the Digambara often married their daughters at ten or twelve. The Shvetambara allowed a second marriage during the wife's lifetime if she had no children, but the Digambara usually refused this. All sects followed the general Indian practice of forbidding the re-marriage of widows, since one who has been free from worldly ties should not seek to form them again. Yet all sects allowed widowers to remarry. Jain women were not traditionally kept in seclusion, but under Muslim influence they adopted forms of purdah as protection and a sign of good social position. [5]

Being barred from many occupations Jain laymen became rich in commerce, and the splendid and lavishly decorated temples of Jainism testify to the generosity of royal and merchant patrons. These lovely temples are surprising in

[4] Uttaradhyayana, 8–9.
[5] See S. Stevenson, *The Heart of Jainism*, 1915, ch. 9.

such a world-renouncing religion and they are found in many parts of India. The most famous are on Mount Abu where the hill is crowned with fine buildings. In other places there are great statues, as at Gwalior where they are carved out of the hillside, and at Shravana Belgola in south India where a great naked male figure, fifty-seven feet in height, rises from an open courtyard. Anthills stand on either side of this huge Jina, and creeping plants are engraved on his legs and thighs, but there is a serene expression on his face.

Of the temples of Khajuraho at least a third are Jain, adorned with rows of sculptured statues. The largest is dedicated to the third Jina, Parshva, and it has many almost naked male and female figures, similar to those of adjoining Hindu temples. Some of these statues show girls putting on adornments, one pulling a thorn out of her foot, and a tender group of a man and woman leaning towards each other, she gazing in adoration while his arm is round her waist and his hand on her full breast.

Jain painting and book illustration had many themes similar to the Hindu. The swastika was a common ancient Indian symbol, said to be derived either from the sun's rays or from the *yoni*, and the magic syllable OM was the nucleus of many Hindu and Jain yantric diagrams. Jain cosmograms depicted the universe with seven separating oceans, interpreted as a cosmic body with OM at its centre. Indian cosmograms, most of which were Jain, illustrated general Indian ideas rather than specifically Tantric. They were intellectual and lacked the passion of Tantra, though they could be used by Tantrists as pictures of human and universal bodies. Jainism accepted some Tantric methods, but never those of the 'left hand'. Tantric teachings from the sixth century, if not earlier, which led to the composition of Hindu and Buddhist texts, influenced Jain iconography and led to the elaboration of their pantheon and rituals.

Jain temples had many images, the central one being a Jina, where there would be a *linga* in a Hindu temple of Shiva. Above this central object rose the spire which represented the world-mountain, Mount Meru. Away from the centre stretched the continents and seas in cosmograms, and the lower slopes were filled with heavenly and earthly beings. In Jain belief men struggle up to the heights through millions of rebirths, but Jains consider that there are multitudes of souls, as against a common Hindu and Tantric view which holds that individual souls are all sparks of one ultimate soul. The individual Jain souls may all attain nirvana, being released from the bonds of Karma, to live in eternal isolation at the top of the universe. Jain icons depict the released spirit as a naked man without any colouring.

The extreme asceticism of the Jains was probably the cause of their small numbers, today less than three millions out of some six hundred million Indians at the latest census. Even the Buddhist Middle Way, which flourished widely in India for over a thousand years, was virtually wiped out by invading Islamic attacks on monasteries and monuments, and even more by renascent Hinduism which swept across India with the passionate appeal of Shiva and Vishnu with his Avatars. The Jains survived, in small numbers, perhaps because there was less dependence on the monastic orders than in Buddhism, and also because of some powerful lay supporters.

SIKH VIRILITY

Independent revival movements stirred central and northern Indian religious life from the fifteenth century, especially among the followers of Kabir and Nanak, though there were many other sects also. Kabir was a low-caste Muslim weaver from Benares, who looked beyond the rival gods and temples of Hindus and Muslims to the one deity whose service was summarized in the letters of the word *prema*, 'love'. Kabir used terms and symbols of Tantra, but he rejected its methods and exercises as contortions and useless.

In popular north Indian mythology the ideal man and woman were the Sur, the Rajput hero who fought unto death on the battlefield in order to fulfil his pledge, and the Sati, his faithful wife, who climbed the funeral pyre in order to be reunited with her husband in death. Kabir used these two figures to express the ideals of perfect loyalty and love to God. But although Kabir was married and had children, his views of sex and family life were pessimistic and bitter. He spoke of woman as 'a pit of hell', 'a black cobra', 'poisonous fruit', 'a blazing fire'.

Some of these harsh words were warnings against liaison with another's wife, which was 'just like eating garlic', for she was 'a sharp blade which few can escape'. But then Kabir turned on sexual relations in general: 'All that man-woman business is really hell, so long as desire persists in the body', because woman ruins everything. 'When a man falls in love with a woman, his wits and intelligence leave him.' Not only adulterers, but the average 'sensual man knows no shame', and 'even the scholar has become shameless.'[6]

Kabir therefore continued the ascetic tradition, and his modern followers number little more than a million. He is very widely respected in India, often compared with the Buddha, admired by Christian missionaries as a monotheist, and revered by the Sikhs who included some of his verses in their scriptures. But his anti-sex verses have often been ignored or under-estimated.

> Woman is the refuse of the world,
>> which sorts out the good and the bad;
> Noble men will put her aside,
>> only the vile will enjoy her.

Guru Nanak, founder of the Sikhs, was a younger contemporary of Kabir but a Hindu of the warrior-ruler caste. Sikhs in India today number over ten millions, mostly in the Punjab where Nanak lived, and there are numerous Sikh communities in Africa and Europe. Nanak was married and had two sons; according to legend they were conceived when Nanak gave his wife two cloves, perhaps phallic symbols. It is said, however, that Nanak 'showed little affection for his wife', and was restless and unhappy, perhaps because of his increasing religious vocation. His wife was often absent, there were family quarrels, Nanak was called 'mad', and eventually he and his wife separated and he became a wandering teacher.

Nanak spoke of God, the One, the Eternal, as all alone before creation. There was no cult of Vishnu then, 'nor that

---

[6] C. Vaudeville, *Kabīr*, 1974, pp. 295 ff.

based on Shiva the passive male, and Shakti the active female, there was neither friendship nor sexual appetite.' Similarly, 'Krishna was not, nor his milkmaids.' Yet in the Sikh scriptures God was called both Vishnu and Shiva, and in popular Sikh art Krishna appeared in his incarnations. Fundamental Hindu beliefs in Karma and rebirth were retained, but mingled with theistic beliefs in grace and salvation.[7]

In the seventeenth century the tenth Guru, Gobind Singh, founded the Khalsa, 'the pure', a militant Sikh company to resist the oppression of Muslim rulers. He summoned all Sikhs to attend a festival of the goddess Durga (Kali) and declared that she required blood sacrifice. In a dramatic initiation he led volunteers into his tent as victims for Durga and came out with his sword dripping blood. This was a test of courage and he revealed that it was goat's blood, dear to Durga, and the volunteers formed the nucleus of the Khalsa. They received baptism or cleansing from water put in an iron vessel and stirred with a dagger, while Gobind's wife, Jita, threw in some sweets. This nectar was sipped by the initiates and sprinkled on them, and all drank from the same vessel, thereby breaking caste taboos in a communion ceremony.

One of the five marks of the members of the Khalsa is wearing long and uncut hair and beard, although modern Sikhs often trim their beards. The wearing of a turban to cover the hair and comb is a cherished tradition rather than a scriptural injunction, and it has caused emigrant Sikhs some troubles. But it is difficult to discover the reasons for the uncut hair, and there have been few Sikh attempts to explain it. Shaving or pulling out the hair, we have seen, was widely practised by Hindu, Jain and Buddhist ascetics, and other holy men wore long matted or unkempt hair. Shaving could be a sign of renunciation of the world, and while some ascetics retained the scalp-lock or tuft on the top of the head, most of them finally cut this off too and threw it away saying, 'I am no-one's and no one is mine.' It is an unexplained curiosity why statues of the Buddha depict him with

[7] T. Singh, etc., ed., *Selections from the Sacred Writings of the Sikhs*, 1960, pp. 104 f.

trimmed but not shorn hair, sometimes in the form of a crown or halo, yet his monkish followers have their heads completely shaven.

In contrast to this ascetic shaving the Sikhs made a symbolic inversion, by commanding initiate Sikhs not to cut hair or beard. This was a symbol of virility, like Samson's hair, and was accompanied by other customs of dress. Hindu ascetics often went naked and smeared with ashes, like Shiva, but Sikhs had to come fully dressed to initiation. Yogis often wore earrings, and these were forbidden to Sikhs. Indian ascetics renounced all ties, whereas the Sikhs affirmed that Gobind Singh was their father and all the Gurus were one. Yogis vowed never to touch weapons, but Sikhs had to wear a dagger to defend their faith and community. [8]

Sikhism, then, became a virile and this-worldly religion, in contrast to much Indian renunciation. The Khalsa was a brotherhood, but open to all classes and both sexes. Rather than abandon home and the world, the Sikh community affirmed that the normal world was its battleground, where its rights were to be defended, by force if necessary. Small wonder that the Sikhs were prominent in the Indian army, favoured by the imperial British, and expert in industry, agriculture, and trade. Their aggressiveness in public demonstrations has provoked protests from Hindus.

In the Khalsa women could be initiated and followed similar duties to men. They were neither secluded nor veiled, and worshipped with men in temples. Women initiates took the name Kaur, 'princess', whereas men were Singh, 'lion'. Some of the nine Gurus who followed Nanak had more than one wife, and the famous nineteenth-century Sikh ruler, Ranjit Singh, had many wives, some married formally and some taken as concubines. He permitted widow-burning among Sikhs, and his tomb in Lahore bears the symbols of four wives and seven concubines who were burnt at his own funeral.

Despite his own unhappy experiences, it is said that Guru Nanak praised women and denounced their oppression,

[8] J. P. Singh Uberoi in *Sikhism*, ed., D. S. Maini, 1969, pp. 125 ff.

saying, 'It is through woman, the despised one, that we are conceived and from her that we are born. It is to woman that we get engaged and then married. She is our lifelong friend and the survival of our race depends on her.'[9] When the Adi Granth scriptures were being compiled it is said that a Hindu poet offered a verse which compared women to sly raiders who carry off men to slay them, but this was rejected.

In 1945 the Sikh shrine committee approved a Rehat Maryada, a guide to the Sikh way of life, which reflected the influence of modern reform movements against Hindu customs which had flourished in Sikhism. Sikhs were told how to have nothing to do with caste, ideas of pollution, full-moon ceremonies, wearing sacred threads, or praying at tombs. Infanticide was condemned outright, as was child-marriage. Other men's wives should be respected as one's own mother, and their daughters as one's own. A man should enjoy his wife's company, and women should be loyal to their husbands. Women should not be veiled, and caste should have no place in marriage.[10]

Social mingling of the sexes has been restricted among Sikhs, even in co-educational schools, except among some westernized groups in cities. Marriage is a family as well as an individual affair, and the wife has to be suitable to the whole family, though Sikhs prefer to speak of assisted rather than arranged marriages. The Sikhs, particularly the initiated Singhs, developed their own form of marriage which recognized women's rights. It was a religious ceremony, and not just a civil contract, and since 1910 this Sikh form of marriage has been accepted by state governments.

The wedding may take place anywhere, but it must be in the presence of the scripture, the Guru Granth Sahib. The couple sit in front of it, and hymns are sung about the nature of marriage and its fusion of two souls into one: 'They are not man and wife who have physical contact only. Only they are truly wedded who have one spirit in two bodies.'

[9] W. O. Cole and P. Singh Sambhi, *The Sikhs*, 1978, p. 142.
[10] Ibid., pp. 168 ff.

PARSI CUSTOMS

The Indian Parsis are an even smaller community, some 120,000, but they follow the ancient religion founded by Zoroaster in Persia, hence their name. They have absorbed some Indian marriage customs and general attitudes, but reject the sexual mythology of Shiva-Shakti or Krishna-Radha, and they look to the supreme and righteous God, Ahura Mazda, for 'the religion of the good life'.

In Zoroastrian myth the first human couple, Mashye and Mashyane, grew from the earth in the form of a rhubarb plant. Like the Vedic and Platonic androgynous primordial beings they were joined to each other, limb to limb. Then they separated out in male and female form, and found themselves in a world made by God but in which evil was active. God warned them: 'You are human beings, the parents of the world, so work in accordance with the right order and mind. Do not worship demons.'[11]

In mythology there was also duality on the supernatural level, between the Wise Lord, Ahura Mazda, and the Evil Mind or Lie, Angra Mainyu or Ahriman. It is not certain whether the myth of the divine twins was original to Zoroaster himself or whether he was reformulating it, but it projected on to the spiritual world the division between goodness and evil which are apparent on earth. The difference, however, was between good and evil, not between spirit and flesh, for in Zoroastrian belief the flesh and material world were good, the creation of God and to be redeemed by destruction of evil. It is important to make this distinction, since the notion of evil matter and flesh infected some other religions, notably the Manichean in Persia, and it influenced later Christianity.

Although it had firm beliefs in life after death, Zoroastrianism was strongly this-worldly and it had no ascetic world-renunciation. In the Vendidad (Videvdat) section of the scriptures Ahura Mazda was made to say: 'I prefer a man with a wife to a bachelor, a man with a family to one without any, and a man with children to one without

---

[11] R. C. Zaehner, *The Dawn and Twilight of Zoroastrianism*, 1961, pp. 42, 267.

children.' And again, 'That place is happy over which a man builds a house with fire, cattle, wife, children, and good followers.'[12] Zoroastrianism therefore rejected celibacy, the monastic life, mendicancy, fasting, and mortification of the flesh.

Parsi religion was not merely good in a negative way but enjoined resistance to evil and active practice of good. Marriage was a good institution, recommended in the scriptures, and it was a merit to help others to marry if they were poor. Despite some Indian customs, Parsi marriage was originally according to Persian tradition in the most solemn part which was conducted by priests. Formerly marriages were arranged by parents, with increasing choice by young people nowadays. Preliminary visits included gifts of silver coins, lighting lamps, presentation of rings, and giving of dowry.

Parsi marriage was celebrated on an auspicious day, still sometimes indicated by a Hindu astrologer, and on the first day a twig of a tree was planted as symbol of fertility. Red marks on the foreheads of the bride and groom were said to symbolize the moon's and sun's rays, but they may also have been fertility signs. There were many ceremonial details, including a cloth curtain which at first separated the bride and groom and then was dropped as a sign of union, and seven threads which tied their hands in an even closer bond. There were blessings in the name of Ahura Mazda, praying for long life and a family of sons and grandsons: 'Learn to do good deeds . . . Keep away from the wives of others . . . Be as fertile as the earth. As the soul is united with the body, so may you be united and friendly with your friends, brothers, wife, and children.'[13]

Chastity is an important Parsi virtue, and husband and wife vowed fidelity and devotion to God and to each other. Adultery was a great evil, opposition to the good spirit, and preventing the progress of the world. In ancient times a husband could divorce his wife for adultery and since the last century, at least, adultery of the husband also was a ground

[12] J. J. Modi, *The Religious Ceremonies and Customs of the Parsees*, 1937 edn., p. 14.
[13] Modi, op. cit., pp. 14 ff.

for divorce. Impotence or sterility could also bring divorce, and until the last century a husband was allowed to take a second wife if the first wife was sterile, but the latter would continue to be mistress of the house.

When a woman was pregnant her husband was not supposed to have intercourse after the fifth month, and she had to abstain from contact with any dead or decomposing matter. For forty days after childbirth the mother was isolated, both in room and bed, and she ate her food separately. After confinement she was purified by taking a bath, administered by a priest with consecrated water. All her bedding and clothes were destroyed or given away to sweepers.

Menstruation was regarded as an unclean thing that was under the influence of Ahriman, the Evil Spirit. Every village or street had a house for menstruating women, and special isolation is still arranged for them. Anything touched by menstruating women became unclean, and if they had their children with them their bodies were washed before being taken out of the house. Sexual intercourse with a menstruating woman received severe penalties. Such women were only given limited amounts of food, to be taken in metal vessels rather than clay or wood, and with the hands covered with gloves. At the end of her period the woman bathed, and washed her clothing and bedding. Most of these practices are still observed, but generally not in separate houses but in private rooms on the upper floors of their homes.[14]

Parsis condemned courtesans as dangerous to both nature and society, their look would dry up the waters, take the bloom from trees and greenness from the earth, and diminish courage, strength, and truth in the righteous. Such women were said to deserve death more than snakes, she-wolves, or brooding frogs. For Zoroastrianism was a religion which faced the perennial problem of evil in the world by seeing life as a battle between the good God and his attendant powers, and the demonic hordes. Human conduct was important because good actions allied men and women with God and worked for the ultimate triumph of good.

[14] Modi, op. cit., pp. 3 ff, 161 ff.

TRIBAL RELIGIONS

There are still many hill and forest tribes in India whose social customs and religions sometimes throw light on Hindu beliefs, and also present differences, while in modern times such people are subjected to increasing pressure from Hindu religion and culture.

Even among the Chenchu, a tribe of jungle nomads in southern India, some of the spirits they believe in have been given Hindu names, such as Bhagavantaru who lives in the sky and controls thunder and rain. Bhagavan is a Hindu name for fortunate, glorious, or adorable, and is applied to many gods as well as to Buddhas and Jinas. The chief Chenchu deity is a goddess, and if a hunter kills a female animal he prays to the goddess for forgiveness. Such deities are pictured in human form, but they have little concern with human morality, even adultery or murder.

The Gond tribes of central India, some three millions, are economically more advanced than the Chenchu and their religion is more complex. Again there is a Bhagavan god who presides over the world, often identified with Shiva, but he receives little ritual worship. More attention is given to clan-deities, most of whom represent an amalgam of male and female, thought of as mother and son but often referred to as a single god. There are also shrines of village mothers, like the mother goddesses of countless Hindu villages, and village guardians who are represented by pointed wooden posts. The earth mother is especially important at times of sowing and harvest. Yet once again it seems that the gods of the Gonds are not much troubled by human morality. Religion is said to be less a personal relationship with invisible beings, than a system of rites and sacrifices whereby human actions are helped by divine influence.[15]

On the other hand, there are many taboos which are traditional, concerned with the safety of society, and sometimes referred to supernatural powers. Pre-marital sex is generally tolerated and adultery is against the social order, but it is not generally thought to incur the anger of the gods,

[15] C. von Fürer-Haimendorf, in *Man and his Gods*, ed., G. Parrinder, 1971, pp. 36 ff.

except in the case of priests and shamans who are channels of divine influence. Such holy men may also have taboos against approaching their wives before or during religious ceremonies in which they are taking part.

The Saora, neighbours of the Gonds, tell a story of a priest who was sleeping in the same house with his wife and his younger brother. The priest was very drunk and slept heavily, and the younger brother took the opportunity to seduce the wife. As they lay together, a sacred pot hanging above them came crashing down and knocked the brother senseless for three days. The wife also developed fever and only recovered after a pig had been sacrificed. The priest himself, however, was not affected.

In marriage many tribal peoples expect fidelity, though priests have a higher standard to maintain than lay people because of the dangerous sacredness of their office. There are taboos on sexual intercourse during menstruation, and priests who break the ban may die. Menstruating women are generally barred from cooking, fetching water, dancing, or sacrificing if they are priestesses. But it is possible that some of these taboos are fairly recent innovations from Hindu influences. Sexual intercourse may be forbidden when there is an earthquake, perhaps because it is believed to be caused when the earth god goes to his wife.

There are taboos on incest everywhere, notably on intercourse with anyone who is related by blood, and it has been thought wrong to marry someone from the same village, so that rules of exogamy apply. Homosexuality and bestiality are religious taboos, which are expected to bring immediate punishment from gods or ancestors. But while boys may imitate the sexual act in playing together, this is generally in the normal man-woman position and shows no fear of breaking a taboo.[16]

Many taboos are confined to women. If sexual intercourse is forbidden while feeding a baby, that is one reason for the husband taking another wife. There are often food taboos for both parents some time after the birth of a child. There are women priests and shamans, but women are often barred

[16] V. Elwin, *The Religion of an Indian Tribe*, 1955, pp. 507 ff.

74

from killing sacrificial animals or burying ashes after cremation. Fear of pollution by woman's blood is the reason for such taboos, as in rituals and bans of female priests in other religions.

The Saora have a story to explain why women lie beneath their husbands in intercourse. Formerly they used to be above while the men lay below. One day two women took a pig to sacrifice to the earth god, and they disagreed about the way they should walk, facing each other with the pig on a pole between them. They quarrelled and the pig's snout fell and went into the genitals of one woman and its tail into the genitals of the other for in those days people were naked. The women threw the pig down and swore they would never again kill animals in sacrifice, or eat pork, and in future they would lie beneath and not above their husbands.[17]

Tribal peoples sometimes live among the ruins of older cultures in the jungle and their customs may throw light on former beliefs. Among the Nagas of Assam there are relics of the capital city of the Kachari kingdom, which flourished till the fifteenth century. In the jungle there remain some fifty huge stone monuments, like an erotic Stonehenge. More than half of these represent the stone *linga* of Shiva, carved in great detail, and between them are great forked stones of the *yoni* of Shakti. The tallest stone phallus is over twenty feet high and five men with outstretched arms could hardly encircle it. Most of the stones are decorated with reliefs of peacocks, parrots, buffaloes, and plants. The meaning of these ancient monuments, unparalleled in the whole of India, becomes clear from the rites of the Nagas themselves, for they set up rough stones in their festivals, and also use carved wooden *lingas* and wooden forked *yonis*. For the ancient Kacharis the *lingas* were evidently memorials of great sacrificial feasts and fertility rites where much blood was shed. The use of stone would guarantee the perpetuation of the fertilizing power of the rituals.[18]

In modern times many of the Indian tribal peoples have been under great pressure, not only from Christian and

[17] Ibid., p. 521.
[18] C. von Fürer-Haimendorf, *The Naked Nagas*, 1962 edn., pp. 27 ff.

Muslim evangelism, but even more from Hinduism to be assimilated into uniform patterns of Hindu culture. Formerly many tribal peoples went about naked or only half-clothed, but now state governments have organized campaigns to clothe them. When Bhil girls in Rajasthan were photographed bare to the waist, the local government issued them with thousands of white saris. Since white is the colour of mourning in India, the result was that these beautiful girls came to look like Hindu widows.

Fuss has been made about tribal people shaking hands, since hand-clasping is a Hindu symbol of marriage, and to touch a member of the opposite sex has been declared degrading to India's high standards of female purity. Verrier Elwin commented that some reformers have devoted their lives to robbing tribal peoples of what little pleasure they have. Turning them into vegetarians has deprived them of essential elements of diet, imposing prohibition has both robbed them of needed tonics and deprived marriages and festivals of their former gaiety, and funerals of some sort of comfort. By insisting on clothing it has been suggested that nudity is indecent, casting a shadow on the delights of love, and taking colour and freedom from life, which is just as bad as other forms of exploitation.[19]

[19] V. Elwin, *The Tribal World of Verrier Elwin*, 1964, p. 336.

Chapter 5

CHINESE YIN AND YANG

FEMALE AND MALE

The duality of the sexes is one of the oldest and most usual ways of representing universal powers and relationships. This symbolism was used in many cultures: Father Heaven and Mother Earth, Zeus and Demeter, Dyaus and Prithivi, Shiva and Shakti, Yang and Yin. But Chinese symbolism developed in particular ways owing to its virtual isolation, and even after the coming of Buddhism in the first century A.D. the new influences tended to be absorbed into indigenous patterns.

In Chinese inscriptions that remain from prehistoric times, pictographs, that were the basis for later writing, represented a woman as a figure with large breasts, and a mother figure had added nipples. The character for man was a square picture of cultivated land with a sign meaning work. These suggest that woman was chiefly the nourishing mother and man the farmer, and that perhaps ancient Chinese society was matriarchal.

Further, the colour red was associated with woman, creative ability, and sexual power, and the marriage ceremony was later called the 'Red Affair', with all the presents and decorations in this auspicious colour. But the colour white indicated sexual weakness and death, and a funeral was the 'White Affair'. In later Chinese alchemical and erotic literature woman was called red and man white, and pictures often showed them in these colours. In ancient times children seem often to have been named after their mothers, and in old myths women had magical powers, while

77

in handbooks of sex women were the teachers of sexual knowledge.[1]

This early matriarchal rule was reversed by the Chou dynasty, which ruled from about 1100 to 221 B.C., and the patriarchal system then imposed was reinforced by the teachings of the Confucians, which emphasized the strength and superiority of man who was the leader and head of the family. Yet powerful counter-currents remained, especially in Taoism, with concepts of the Great Mother, and the potent female who in sexual intercourse fed man's limited life-force from her inexhaustible supply. The negative was praised above the positive, and inactivity above activity. In classical Taoist texts mystical terms like 'the deep Valley' and 'the mysterious Doorway' were interpreted in sexological texts as womb and vulva. A famous chapter of the classical Taoist scripture, the Tao Te Ching, could be taken in this symbolical manner:

> The Valley Spirit never dies.
> It is named the Mysterious Female.
> And the Doorway of the Mysterious Female
> Is the base from which Heaven and Earth sprang.[2]

The alternation of day and night, summer and winter, youth and age, led the Chinese to believe in the interaction of dual cosmic forces. Mankind functioned in the same way as the universe, and human sexual intercourse was like the union of heaven and earth which mate during rainstorms. The Chinese thought of the clouds as the ova of the earth, which were fertilized by the sperm of heaven in rain. Much of the symbolism of sun and moon, or heaven and earth, came to be replaced by the formal terms of Yang and Yin, but the symbol of the mating of heaven and earth in storms remained, and to this day 'clouds and rain' is a standard expression for sexual intercourse.

A classical story expressed this idea in an account of a king who went on an excursion and fell asleep during the daytime. He dreamt of a woman who said, 'I am the Lady of the Wu

---

[1] R. H. van Gulik, *Sexual Life in Ancient China*, 1961, pp. 5 ff.
[2] A. Waley, *The Way and its Power*, 1934, p. 149.

mountain and wish to share pillow and couch with you.'
They had sexual intercourse and on parting she said, 'I live
on the southern slope of the mountain. In the morning I am
the clouds and in the evening I am the rain.' The story
transformed the old picture of the mating of heaven and
earth, and it was the woman who made the sexual advances.

In addition to 'clouds and rain', later literature wrote of
'the Wu mountain' or 'the southern slopes of the Wu
mountain', as elegant terms for coition. In sexological
writing the 'clouds' were explained as both ova and vaginal
secretions, and 'rain' as semen. Expressions such as 'the
reverse clouds and the inverted rain' were used of male
homosexual acts.[3]

YIN AND YANG

The terms Yin and Yang appeared in Chinese philosophy
from the fourth century B.C., though archaeologists trace
their symbolism to a much earlier time. The origins of the
characters for Yin and Yang is not known, but they were
interpreted as 'the dark side' and 'the sunny side' of a hill,
and from that there developed the indication of vital
energies: dark and light, weak and strong, female and male.
The female was therefore dark, black, deep, and receptive;
and the male was bright, high, celestial, and penetrating. Yet
they were complementary rather than opposing, since all Yin
had some Yang in it, and all Yang some Yin.

Yin and Yang have been compared to the Dark and Light
of Zoroastrianism, and the dualism of evil and good that
arose therefrom. But Yin and Yang were not opposed, they
were interdependent, like woman and man. The aim was not
the triumph of one over the other, but a perfect balance of
the two principles.

Later Chinese philosophers gave much time to speculation
about the beginning of things. The ultimate unity, as in
Indian, Persian, and Greek notions of the undivided male
and female, was represented in China by the symbol of a
circle. From the eleventh century A.D. Neo-Confucian

[3] Van Gulik, op. cit., pp. 38 f.

scholars represented this concept by the undivided circle known as *t'ai chi t'u*, 'the supreme ultimate'. The circle was divided into two pear-shaped halves of dark and light, Yin and Yang. The dark half Yin contained a white dot indicating the Yang embryo within it, and the light half Yang had a black dot designating the Yin element in it.

Although the philosophical interpretation came late, the design itself went back to ancient times with circles found on ancient bronzes. It remains to this day in decorations on gates and houses, utensils and furnishings, and in sexual and exorcist symbols. The circle with its pear-shaped halves is also found in India and Europe, and it appears in patterns popularly known as Kashmir or Paisley.

Different and more complex magical symbols were developed in China in combinations of horizontal lines, which were interpreted in books of divination. Among these the *I Ching*, the *Book of Changes*, came to supersede all others. This book became so important in daily life that it was considered as a 'Confucian' classic, though it was not adopted by followers of Confucius until well after his time. Tradition credited a mythical emperor with the invention of eight basic trigrams of horizontal lines; broken lines representing Yin and continuous lines being Yang. The eight trigrams were again combined into pairs and formed sixty-four hexagrams. Arranged in a circle, symbol of heaven, the trigrams corresponded to the directions of the compass, the seasons of the year and the times of day.

The *I Ching* described Yin and Yang as the dual cosmic forces that perpetuate the universe in a chain of permutations, a concept that was worked out into a philosophical system, so that the *I Ching* was used by both philosophers and diviners. Here we are only concerned with passages that consider the relationship of the sexes. Thus it was said that 'the constant intermingling of Heaven and Earth gives shape to all things, and the sexual union of man and woman gives life to all things.' Again, 'the interaction of one Yin and one Yang is called Tao', and this was later interpreted to mean one woman and one man.

Hexagram 63 was considered to symbolize sexual union, for it consisted of the trigram meaning 'water', 'clouds', and 'woman' on top, and beneath it the trigram meaning 'fire', 'light', and 'man'. Thus the harmony of woman and man was expressed and their complementary nature was depicted by the perfect alternation of Yin and Yang lines, and such harmony was regarded as the basis of a happy sex life. 'The transition from confusion to order is completed, and everything is in its proper place even in particulars. The strong lines are in the strong places, the weak lines in the weak places. This is a very favourable outlook.'[4]

Nearly all later Chinese handbooks of sex speculated on this hexagram and pictures showed scholars meditating on the perfect balance of the male and female elements here. It is significant that the element for woman occupied the upper part, and as to the symbols of fire and water medical treatises said that man was like fire which flared up easily but was extinguished by female water which heated and cooled slowly, as in human sexual experience.

TAO

Theories developed from the *I Ching* influenced Chinese ideas on sexual intercourse, but with the rise of Taoism in the second half of the Chou dynasty much wider currents of thought came to the fore. The concept of Tao was basic to much of Chinese thought and art, religion and sex. This complex word has many shades of meaning: way, power, principle, and it was beyond meaning: 'The Tao that can be defined is not the ultimate Tao, it existed before Heaven and Earth, it is the ancestor of all doctrines, the mystery of mysteries.'

Philosophical Taoism, exemplified in the classic Tao Te Ching, spoke of Tao as like an empty vessel, a mysterious female, an uncarved block, water which takes the lowest way but benefits all creatures. The sage similarly overcame by 'wordless teaching, actionless activity, discarding formal knowledge and morality.' It seemed likely that a man who followed nature and did what he wanted would enjoy sex

---

[4] *The I Ching or Book of Changes*, tr. R. Wilhelm and C. F. Baynes, 1951, i, p. 260.

without limitations, unless he was seeking a higher stage of trance. Perhaps this is what was meant by an obscure passage: 'An infant has power, its bones are soft, but its grip is strong. It does not yet know the union of male and female, but it may have an erection and show the height of vital force, which means that harmony is at its perfection.' So the sage may use continence like a child, to help him concentrate his energies and attain a tranquil state. [5]

Popular Taoism was the religion of the masses in ·China, and its symbolism appeared in mythology, art, medicine, magic, and sex. Common symbols of Tao were twisted or hollow stones, which were sought by collectors from rivers and lakes, and kept in gardens and houses. The stone could represent both the kidney shape of the female Yin vulva, and the mountain shape of the male phallic Yang in harmony. In carvings gods and pilgrims going to their temples could be fashioned inside the same stones. Taoist ideas brought objects into contact with each other, or arranged them to influence one another, so that the aroused Yin and Yang would be in balance.

Yin and Yang symbolism was seen by Taoists in countless objects, patterns, pictures, vessels, public and domestic altars. Yang was thought to dominate in stallion, dragon, Feng-bird, cock, ram, horned animals, mountains, summer, and the south. Yin dominated in fungus, whirling clouds, water, valleys, winter, north, vase, peach, female dragon with divided tail, fish, peony, and chrysanthemum. The peach, with its deep cleft, was a favourite symbol of the vulva. Harmonic combinations showed a Feng-bird flying into a garden of peony flowers, a dragon among swirling clouds, a woman with a ram, or a plate with subdued red and blue flowers melting together to show the reconciliation of colour opposites. Chinese art was imbued throughout with mystical and sexual symbolism, and to understand it is to learn an unspoken language. [6]

Popular Taoist myths used sexual symbolism in accounts of the gods. A favourite deity was Hsi Wang Mu, the Royal

[5] Tao Te Ching, 2 and 55; H. Welch, *The Parting of the Way*, 1957, p. 71.

[6] P. Rawson and L. Legeza, *Tao*, 1973, pp. 12 ff, and plates 51, 62.

Mother of the Western Air, or the Golden Mother of the Tortoise, the counterpart of the god Mu Kung who was sovereign of the Eastern Air. She was the passive or female Yin to his Yang, and by their union all beings in heaven and earth were born. Hsi Wang Mu lived on a jade mountain in a palace with a Heavenly Peach Garden and a magic tree that ripened every six thousand years. Then it was the birthday of the goddess and all the immortals were invited to the Peach Festival, at which they ate the fruit that gave them immortality. A famous Buddhist story of 'Monkey' told how he stole the food and pills of immortality when he gate-crashed this festival. Hsi Wang Mu was depicted in art in brilliant dress, accompanied by a phoenix and bearing a dish of fruit. In other pictures the immortals appeared outside the golden rampart of her palace hoping to enter, while others were already seated on an island in her garden lake feasting on the immortal peaches.

Among other gods the star deity Shou Hsing or Shou Lao was the prototype of male success. As the Old Man of the South Pole, his influence brought peace to mankind. In pictures and carvings he was characterized by a very high forehead and bald pate, like a mountain to show his great power, and a peach which he carried in his hand indicated long life. He was a happy god, carrying a long staff to which were attached a gourd and a scroll which with the peach were symbols of longevity.

There were countless Taoist gods and divinized human beings, each with individuality and eccentricity. Legendary immortals were said to ride through the air, on dragons or in chariots, and some lived in caves on mountains where they sought the food of immortality and passed on their discoveries to favourite pupils. Most of their images and temples have disappeared, outwardly at least in China, though traces survive in other places of Chinese influence, and in works of art in galleries and museums all over the world.

The search for long life or immortality was a constant aim of Taoists. They practised breathing exercises, like Indian yogis, but making breathing quiet and holding it as long as possible, with the aim of returning to the manner of respiration in the womb. They practised sun-bathing (only recently recognized as valuable in the West), while holding in the hand a character of the sun within a border. But women had to expose themselves to the moon, holding a piece of yellow paper with the moon in a black border. Those who sought the 'outer elixir' were alchemists looking for magical potions, drinking dew, hanging from trees, returning to nature, or setting off for the islands of the immortals. In such a search a Taoist might abandon sex altogether, or restrict orgasm so as not to lose the precious fluid. But those who pursued the 'inner elixir' tried by self-control and proper techniques to achieve perfect co-ordination of body and soul.

TAO IN SEX

'Of all the ten thousand things created by Heaven, man is the most precious. Of all things that make man prosper none can be compared to sexual intercourse. It is modelled after Heaven and takes its pattern by Earth, it regulates Yin and rules Yang. Those who understand its significance can nurture their nature and prolong their years; those who miss its true meaning will harm themselves and die before their time.' So wrote the scholar Tung-hsüan, perhaps in the seventh century A.D. in his *Art of Love*. [7]

From the highest to the lowest levels of Chinese society followers of Tao sought to cultivate sexual energy and unite Yin and Yang. In sexual play these powers were aroused and in orgasm they were released from the body and passed into the partner of the other sex, male into female and female into male. Such mutual exchange of Yin and Yang essences was thought to produce perfect harmony, and sexual intercourse, instead of declining with age, was believed to increase vigour and bring long life.

Tung-hsüan said again:

Truly Heaven revolves to the left and Earth revolves to the right. Thus the four seasons succeed each other, man calls and woman follows, above there is action and below

[7] R. H. van Gulik, *Sexual Life in Ancient China*, pp. 125 ff.

compliance; this is the natural order of all things . . . Man and woman must move according to their cosmic orientation, the man should thrust from above and the woman receive below. If they unite in this way, it can be called Heaven and Earth in even balance.

Sexual intercourse in China had two main purposes. The first was the procreation of children, especially healthy males, when the man's Yang essence was at its highest strength, so that the family would be continued, the ancestors cared for, and the order of the universe maintained. But male semen was supposed to be strictly limited, while woman had an inexhaustible supply of Yin essence.

The second purpose therefore was to strengthen male vitality by absorption of female Yin essence. The Yin essence was supposed to be in the vaginal juices which the man absorbed, but a further element was added by *coitus reservatus*, coition without ejaculation, so that the Yang would be supplemented by the Yin. If this was done with several partners, prolonging coitus as much as possible without orgasm, then the Yang would be augmented and strengthened. Although Taoism stressed the importance of the co-operation of Yin and Yang, this did not necessarily lead to love and equality between two partners. In rich families, at least, there were numerous concubines, and since the dominant male could not often ejaculate every time, limitation of emission helped to solve the problem, with the belief that retained semen aided health.

Some of the sexual techniques of the Taoists were greatly opposed by Confucians and Buddhists. Taoism considered continence to be against the rhythm of nature, and celibacy to lead to neurosis, whereas the Buddhists advocated monastic celibacy. Further, Taoists taught not only *coitus reservatus* by mental discipline, but by physical methods. Ejaculation was prevented by pressing the seminal duct with the fingers, thus diverting the fluid into the bladder. But Taoist theory, like Indian Yogic and Tantric, held that the semen (*ching*) would 'flow upwards' along the spinal column to 'nourish the brain' and the entire system, the male Yang essence having been intensified by contact with the female Yin.

Chinese literature on sex emphasized that semen was a man's most precious possession, and every emission must be compensated by acquiring an equivalent amount of Yin essence from the woman. While he should give the woman complete satisfaction at every act of coition, he should only ejaculate on certain occasions. If they wished the woman to conceive then the most favourable time was thought to be five days after menstruation.[8]

Reprehensible Taoist practices, in the eyes of other religions, were public ceremonies of sexual intercourse which flourished in the early Christian centuries and resembled Indian left-handed Tantra. After a liturgical dance the two chief celebrants might copulate in the presence of the congregation, or members would do the same in chambers along the sides of the temple courtyard. Sexual intercourse was sacralized, and the human union harmonized with that of the universe, with careful attention to the seasons, the weather, the phases of the moon, and the astrological situation.[9]

Sexual handbooks were compiled from ancient times in China, often written as dialogues. The mythical Yellow Emperor, Huang-ti, figured largely in handbooks from the early centuries asking questions of female guardians of the mysteries of sex. One of the chief of these was the Plain Girl, Su-nü, said to have been a river-goddess who could take the shape of a shell, a fertility symbol in China. A story is told of a poor but virtuous man who found a large shell on a river bank and took it home, whereupon every time he went out he found the house cleaned and food prepared on his return. He watched secretly and saw a beautiful young girl emerge from the shell, who said she was Su-nü sent to look after him by the Heavenly Emperor. She disappeared, but the shell remained and was always full of rice.[10]

The Plain Girl was said to have explained the arts of sexual intercourse, along with two others, the Dark Girl and the Elected Girl. The Elected Girl is a rather nebulous figure, though she is also said to have been a goddess. But it was

[8] R. H. van Gulik, *Sexual Life in Ancient China*, pp. 46 ff.

[9] J. Needham, *Science and Civilisation in China*, 1956, ii, pp. 146 ff.

[10] R. H. van Gulik, *Sexual Life in Ancient China*, pp. 74 ff.

reported that, 'The Yellow Emperor learned the Art of the Bedchamber from the Dark Girl. It consists of suppressing emissions, absorbing the woman's fluid, and making the semen return to strengthen the brain, thereby to obtain longevity.'

It is clear that in the early centuries there existed a number of handbooks of sexual relations, written as dialogues between the Yellow Emperor and one of these Girls. These manuals were illustrated with pictures of various positions of coitus, their methods were widely known by husbands, wives, and dancing girls, and formed part of a bride's trousseau. They not only taught how to maintain satisfactory sexual relations, but also how to benefit health and prolong life.

In the second century a well known poet Chang Heng described a bride addressing her husband. She swept the pillow and bedmat, lit the lamp and filled the burner with incense. She locked the door, shed her robes, and rolled out the picture scroll to take the Plain Girl as her instructor as taught to the Yellow Emperor, 'so that we can practise all the variegated postures, those that an ordinary husband has but rarely seen.'[11]

Master Tung-hsüan, quoted at the beginning of this section, said, 'The methods of sexual intercourse as taught by the Dark Girl have been transmitted since antiquity; but they give only a general survey of this subject, they do not exhaust its subtle mysteries.' He gave many more details therefore of embracing, kissing, petting, licking, biting, thrusting, intervals, times, and seasons.

Thirty main positions were described for sexual intercourse, each with a metaphorical name: Unicorn's Horn, Winding Dragon, Pair of Swallows, Fluttering Butterflies, Reversed Flying Ducks, Bamboos near the Altar, Galloping Steed, Jumping White Tiger, Phoenix in a Cinnabar crevice, and so on.[12] Here and in other writings descriptive names were given to the sexual organs. The male was the Jade Stalk, the Positive Peak, the Swelling Mushroom, the Turtle Head

---

[11] R. H. van Gulik, *Sexual Life in Ancient China*, p. 73.
[12] Ibid., pp. 125 ff., much detail in Latin.

or the Dragon Pillar. The female was the Jade Gate, the Open Peony, the Golden Lotus, the Receptive Vase, the Lute Strings, the Golden Gully, the Deep Vale, the Chicken's Tongue. The most powerful symbolical natural substance was Cinnabar, a rosy-purple crystalline stone, which represented the energies of joined Yin and Yang.

Handbooks of sex continued to be popular in China with the centuries. Sometimes numerous items on the Art of the Bedchamber were listed under Taoist Classics, and lists of medical books also included works on sex, such as, *Classic of the Secret Methods of the Plain Girl*, *Handbook of Sex of the Dark Girl*, *Summary of the Secrets of the Bedchamber*, *Principles of Nurturing Life*, *Poetical Essay on the Supreme Joy*. By the time of the T'ang dynasty, from the seventh to the ninth centuries, sexual instruction was classified as a branch of medical science, and many handbooks had a special section on it.

The most famous of these medical works was *Priceless Recipes* by the seventh-century physician Sun Szu-mo. In a section on 'healthy sex life', Sun wrote that after his fortieth year a man's potency decreased and he needed to acquire a knowledge of the Art of the Bedchamber. Sexual techniques should be learnt because in youth man does not understand Tao, and in old age he may be too weak or sick to benefit by it. *Si jeunesse savait, si vieillesse pouvait.* Sexual intercourse should not be indulged in simply to satisfy lust, but it should be controlled to nurture the vital essence.

Sun pointed out the importance of preliminary sex play, rousing the woman's passion and drinking her Jade Fluid, her saliva. It did not matter if the woman was beautiful, as long as she was young, but the ideal was to copulate on one night with ten different women, without emitting semen once. Women should be changed, since by intercourse with one only her Yin essence would become weak and of little benefit to her partner. Ejaculation should be controlled by holding the breath and pressing the urethra, so that semen would ascend to the brain 'thus lengthening one's span of life.'

Sun told a story which was often quoted by others. A peasant of over seventy told him that his Yang was exuberant and he made love to his wife several times a day. Sun replied that this was most unfortunate, it was a last flare-up of fire, and the peasant should have abstained from intercourse long ago. Six weeks later the peasant died, and this was a warning to control sexual relations.

'Man cannot do without woman', said Sun, 'and woman cannot do without man', but the passions should not be indulged freely for that would rob the man of his vital essence. Men should nurture their vital power when they notice that it is particularly strong, and in advancing years if a man restrained himself this was like adding oil to a lamp that was about to go out. A strong man of over sixty may feel that his thoughts are still composed after not having copulated with a woman for a month or so, and if he can control himself for so long then he could continue longer. The dominant aim was to conserve the precious semen and prolong life, and the feelings of the woman or women were of less concern, though their Yin essence would be stimulated by coition, even without emission.[13]

CONFUCIAN MORALITY

Taoists, Confucians, and Buddhists have been followers of three traditional 'ways' in China. These were not separate 'religions', and even when opposed they often supplemented each other. The Confucian scholar-gentleman was depicted as a conscientious official, a responsible citizen, and a good family man. The Taoist was often the same man in private life, with his loves and his search for the things of the spirit.

Taoist legends spoke of the abasement of Confucius before their mythical founder Lao Tzu, but they were obviously partisan. Both sages, or the writings attributed to them, spoke of Tao but in different ways. Many Chinese accepted the treatments of Tao as complementary, but some Confucians attacked Taoist teachings as narrow, making light of humanity and righteousness. Taoists had said that robbers would not disappear until the accepted sages died off, but this was thoughtless, for if there had been no sages

---

[13] R. H. van Gulik, *Sexual Life in Ancient China*, pp. 193 ff.

in the past morality would have perished. Taoists and Buddhists had taught men to reject order, and seek for personal purity and nirvana. But to regulate families one must first cultivate one's own person.

Confucius said little about women and nothing of physical sexual relationships. In one verse he remarked: 'Women and people of low birth are very hard to deal with. If you are friendly with them, they get out of hand, and if you keep your distance, they resent it.'[14] The followers of Confucius developed this attitude to ensure the lower place of women; their foremost duty was to obey their husband and his parents, to look after the house, and to bear healthy male children. Procreation was primary, and enjoyment of sex secondary; the ideal woman was 'she who is within', concentrating on household tasks. Chastity was essential for the woman, but not for the man.

Confucian teaching was that of the Mean, the moral ideal of balance in society and harmony with the universe. This was to be practised in the Five Relationships which concern everybody; the relations of ruler and subject, father and son, husband and wife, older and younger brother, and friend with friend.

The Confucians advocated the separation of the sexes, in order to maintain the purity of family life. The *Book of Rites*, a collection of early and later dates, took this separation to extremes: 'In the dwelling house, outside and inside are clearly divided; the man lives in the outer, the women in the inner apartments. The latter are located at the back of the house, the doors are kept locked and guarded by eunuchs.' Husband and wife should not even use the same clothes-rack, they should not bathe together, share the same sleeping-mat, or borrow each other's articles of dress. If the husband was absent the wife should lock his bedding away. She should not receive anything from another man directly. 'When a woman goes out she shall veil her face . . . Walking in the street the men shall keep to the right, the women to the left.'[15]

[14] *Analects* 17, 25, tr. A. Waley, 1938.
[15] R. H. van Gulik, *Sexual Life in Ancient China*, pp. 58 ff.

In the first and second centuries A.D., Lady Pan Chao, greatly honoured for her chastity and learning, advocated elementary education for girls as well as boys. But she wrote *Women's Precepts*, which has been called 'one of the most bigoted books in Chinese literature', though Confucians took it as a shining example of womanhood. Education, thought Lady Pan, should teach woman her inferiority and absolute obedience to her husband.

> The Tao of husband and wife represents the harmonious blending of Yin and Yang, it establishes man's communion with the spirits, it reaffirms the vast significance of Heaven and Earth, and the great order of human relationships . . . Yin and Yang are fundamentally different, hence man and woman differ in behaviour. Strength is the virtue of Yang, yielding constitutes the use of Yin. Man is honoured for his power, woman is praised for her weakness . . . To be reverent and obedient, that is the golden rule of wifehood.

And again, 'According to the *Rites* man has the right to marry more than one wife, but woman shall not follow two masters . . . Modesty is the cornerstone of virtue, obedience the proper conduct of a wife.'[16]

Lady Pan's precepts were ideal, but it seems that they were not often followed. Other writers speak of women going out to pleasures, by day and night, visiting Buddhist temples for festivals and organizing picnics. In the home visitors would tease their host until he brought out his womenfolk, and then they would sit together, sing and dance, and exchange unseemly conversation.

Confucian teachings determined the place of man and woman in society and the family, but in the privacy of the bedchamber they would follow Taoist teachings, and there the woman was not infrequently the teacher of sexual mysteries. Physical contact of husband and wife was confined to the marriage bed, which was often a small room. Even here the Lady Pan maintained that 'Dalliance in the bedchamber will only create lewdness; lewdness will induce

[16] Ibid., pp. 97 ff.

91

idle talk; idle talk will generate moral laxity; and moral laxity will breed contempt for her husband in the wife. The root of all these evils is their inability to learn moderation [in their sexual relations].' It must be said that Lady Pan had married when she was fourteen, but her husband died young and she never remarried though she lived to a great age.

Confucians, like Taoists, and they were often the same persons, thought that sexual intercourse was good, necessary for all men and women, and essential to the continuation of the race. Some teachers disliked sexual dalliance out of fear that it might disrupt family life, and especially hinder procreation by superfluous amorous play. They considered woman to be inferior to man, just as Earth was inferior to Heaven, but they did not despise women or sex, as did teachers in some world-denying religions.

According to Confucian ideas, a man's interest in his wife as a human being ceased when she left his bed. Since women were not supposed to share their husband's intellectual interests, or interfere in their outside activities, little was done for the education of girls. Most women were illiterate, even in upper-class families where they were only taught sewing and weaving. Only courtesans and singing girls learnt to read and write as part of their training.

In Confucian China the relations of the sexes were formal in public. An upright man would show no sign of intimacy with any woman before others, not even with his wife, for that would be detrimental to filial piety, since the primary duty in life was towards the parents. The mother-in-law problem was extended to other relatives of her husband's. Lady Pan said: 'How must a wife gain the affection of her parents-in-law? There is no other way than complete obedience. If her mother-in-law says "It is not" while it is so, the wife must still obey her . . . And in order to obtain the affection of her parents-in-law, she must first secure that of her brothers- and sisters-in-law.'

For a woman to be attractive to men was regarded as unnatural, equivalent to sexual offence, and even attractiveness to the husband should not be displayed publicly. It was good manners to avoid talking to a husband intimately in public, and tender feelings were banned from open company. On the other hand, the husband might, and often did, take a concubine. Very often this would be if the wife was barren, and she might even encourage him to take such a woman to continue the family line. Or a concubine might be acquired if a bridegroom did not like the bride whom he had first met at the wedding ceremony.

Concubinage was an accepted custom in China, though no woman wanted to be a concubine. Girls were sold for money, or sometimes because they had lost their virginity, but even poor families did not like to admit that their daughter might become a concubine and it was said that concubines came from some distant region. A concubine might be well treated if the wife was not jealous or the husband was strong-minded, but her life would be miserable if that was not so. If she failed to give birth to a son she might come to be ignored, while the husband would get other women to ensure the family succession. Since they were often young girls, concubines tended to outlive their husbands and then the family might neglect or scorn them and they would have no relatives to fall back on.

For the husband, possession of concubines or several wives meant increased sexual activity; he would go from one to other of them, on the same or successive nights. On Taoist theory his Yang would be weakened by loss of semen, but by practising *coitus reservatus* he was supposed to satisfy the sexual needs of his wives and concubines, and be strengthened by the passage of their Yin essence to him.

MARRIAGE

The *Book of Rites* attributed to Confucius the saying, 'If Heaven and Earth were not mated, the ten thousand things would not have been born. It is by means of the great rite of marriage that mankind subsists throughout the myriad generations.' Thus marriage was extolled, even when the wife was regarded as inferior. Every woman, however poor or ugly, could claim the right to a husband, and it was the duty of a householder to provide husbands for all the women in his employ; while among the poor and peasants it was the obligation of the community to find a husband for every girl.

In early China marriage among the ruling classes was exogamic, and marrying a woman with the same surname was completely taboo. Peasants also had taboos, though many were not recorded, but the family-name taboo still applies today. Marriages were arranged by a go-between, and the couple did not meet until marriage presents had been exchanged, the selection being decided by the parents. In country communities, in ancient times, young men and girls met at spring festivals, danced and sang together often in erotic ways, and then had sexual intercourse with chosen partners. Such unions would be regularized by the community. The Confucians were shocked by such traditional matings, and ordered that all unions should be supervised and registered by a 'middle-man', though how far this was done is debated.

There is frequent reference in later literature to a custom called 'making a row in the bridal chamber' or 'ragging the bride'. After the wedding banquet the guests would take the couple to the bridal chamber and tease and make fun of them without restraint. Coarse questions and offensive behaviour, sometimes even with whipping, were applied to the 'happy' couple by drunken guests. To some degree the custom subsisted until modern times, perhaps preserving the ancient purpose of defloration of the bride and consummation of the marriage.

The preliminary viewing of the future pair traditionally took place after family agreement, though from the time of the Ming dynasty this viewing did not happen until the bride's veil was removed in the ceremony in the ancestral hall. Courtesans often acted as matchmakers, and they would lead the bride to the nuptial room, where the couple exchanged a wedding-cup and locks of their hair were knotted together.

In ancient China a wife as well as a husband could ask for a divorce and she did not thereby surrender her independence, but Confucians in the Sung dynasty forbade any female remarriage, for 'to die in starvation is a minor matter, but to lose one's chastity by remarriage is very serious'. The *Book of Rites* had already forbidden widows to wail at night, and it even prohibited people making friends with the sons of widows. For the sake of Confucian decorum young women were forced to remain widows, had no freedom, and lived restricted lives in ladies' chambers that came to be called locked-up prisons.

Until the T'ang dynasty Chinese women exposed their throats and bosoms, and girls often danced with naked breasts. But from the Sung dynasty, from the tenth century, the high collar became the distinctive feature of women's dress that it remained till modern times, and communist boilersuits perpetuate this covering. The binding of women's feet, introduced in the tenth century, designed to make women more attractive, was the cause of great suffering and has been abolished.

BUDDHIST INFLUENCES

According to tradition Buddhism, in its all-embracing Mahayana form, entered China in the first Christian century. Although Mahayana could easily absorb native Chinese deities under the guise of Bodhisattvas, 'beings of enlightenment', some Buddhist beliefs and practices aroused criticism as alien to China. The foreign origins of Buddhism were objectionable, as was the claimed superiority of monks to rulers and even to the emperor. The celibacy of monks was one of the greatest obstacles to the native adoption of the new religion, since childlessness was regarded in China as the most unfilial conduct. Buddhist apologists replied that 'Wives, children and property are the luxuries of the world . . . but the monk accumulates goodness and wisdom instead.'

Buddhists found that the Chinese were interested in magical spells and charms, such as were already used in Taoism, and they translated Indian works which included these and teachings about sex. Moreover women had a leading role in Indian sex books, often being the instructors in sexual mysteries, so that Buddhism enhanced the position of women, and joined in this with Taoism generally against the Confucian subordination of women. In the early periods Buddhist texts on love-making were abbreviated to spare Confucian feelings, but when Buddhism was flourishing erotic Tantric texts were widely published. Then under later Neo-Confucian reformers they were expurgated again.

Women, in particular, were attracted by Buddhism. Its creed of universal compassion implied the equality of all beings and answered the spiritual needs of women. The favourite Mahayana scripture was the Lotus Sutra which presented the Bodhisattva Avalokiteshvara, the 'lord of looking or regard', who became in China the compassionate goddess Kwanyin, the lady giver of children (in Japan she became Kwannon or Kannon). Kwanyin was often depicted carrying the peach of sex, she gave children and helped in distress, and the dazzling ceremonies of her temples brought colour into the monotony of daily life.

The most popular Buddhist sect was the Pure Land, the religion of faith and grace. It was ruled over by Amitabha, the Buddha of Boundless Light in his Western Paradise which could be entered by anyone who uttered his name in sincere devotion. The Chinese version of Amitabha was O-mi-t'o-fo, which became a favourite exclamation of astonishment or delight for Chinese women. The Pure Land has remained the most popular of Buddhist sects in Japan, and was for long so in China.

Buddhist nuns were popular among many Chinese women, for they had free access to the women's quarters of houses and became advisors on personal problems. They held prayer-meetings for recovery of sick children or curing of sterility, and for healing female diseases. Nuns also taught girls to read and write, and other feminine skills. But Confucians were shocked at the sight of women forsaking their sacred duty of producing children and satisfying their husbands, and scurrilous novels were written about monks and nuns. Poems were written against monks who were charged with having changed the Pure Land into a sea of lust, like leeches sucking blood; they called a maiden to feel for an opening and revealed the true shape of the Buddha's Tooth.

Some Chinese girls became nuns no doubt for sincere reasons of religious conviction, while others were vowed by their parents as protection against evil, and other wives and concubines escaped to nunneries from cruel husbands or families. Critics said that they practised unnatural vices in the monasteries and nunneries, or they provided love philtres for women and acted as go-betweens in illicit affairs. If nunneries were ruled by women of strong personality the discipline was probably strict, but laxity might give room for suspicion. An erotic text of the T'ang dynasty accused nuns of having sexual intercourse with Chinese and Indian monks, and 'when they are with those lovers the nuns forget the Law of the Buddha and play absent-mindedly with their rosaries.'[17] Others maintained that nunneries were havens for loose women who did not wish to register as prostitutes, but they held dinners and drinking bouts there while the religious authorities made good profits out of the food and wine.

Mahayana Buddhist philosophical systems included speculations about male and female cosmic principles, which were developed in Tantra and resembled Yin and Yang. In the Vajrayana, the 'thunder bolt vehicle', in which the thunderbolt was a phallic symbol, there was taught the attainment of supreme bliss by the union of male and female, a mystical marriage which overcame sexual duality in a hermaphroditic unity. Buddhist Tantrists practised this method either alone in imagination or in real sexual embrace with a woman, but most texts stated that the partner should be a live woman because 'Buddha-hood abides in the *yoni*'. The yogic method of breath-control worked on the man's rising semen so that it was not emitted but rose upwards, and the woman's energy stimulated and blended with his, to produce from the unshed semen a new and powerful essence which rose up through the nerve centres to the 'Lotus on top of the head'.

Taoist sexual practices were no doubt influenced by Indian Tantrism, but it is possible that Indian texts in their turn were affected by Chinese teachings. Buddhist Tantrism arrived in China relatively late, about the eighth century, and since a sexual mysticism of making 'translated semen return' had already flourished in China for several centuries, and had perhaps not been long known in India, there may have

[17] R. H. van Gulik, *Sexual Life in Ancient China*, pp. 175, 206, 266.

been a two-way traffic of religious and sexual practices. An Indian Tantric text, Rudra-yamala, told of a sage who practised austerities for ages without seeing the supreme goddess, whereupon he was advised to obtain the 'Chinese discipline' in which the goddess delighted. She sent him to China where he saw the Buddha surrounded by naked adepts who drank wine, ate meat, and engaged in sexual intercourse with beautiful women. The sage was greatly disturbed by this sight, until the Buddha taught him the true meaning of the sexual rites.

Vajrayana became practically extinct in India by the twelfth century, but its teachings had been re-imported into Tibet and China. There was opposition to it, and a shocked Confucian scholar in the Sung dynasty described a 'Buddha-mother Hall' in Peking in which Tibetan male and female deities were represented by statues locked in sexual embrace. These were regarded as repulsive foreign practices, without any recognition that they might have depicted ancient Taoist disciplines. Earlier such 'Joyful Buddhas' had been used to instruct princes and princesses in sexual matters, but Neo-Confucianism tried to suppress 'immoral cults', and description of the sexual mysticism of Chinese Tantrism was partly preserved in Japan.[18]

VARIATIONS

Since the Chinese believed that a man's semen was his most precious possession, and every emission would diminish his vital force unless compensated by acquiring an equivalent amount of Yin essence from a woman, it followed that certain sexual variants were reprehended. Male masturbation was forbidden, for it implied a complete loss of vital essence, and it was only condoned when a man was away for long from female company and the 'devitalized semen' might clog his system. Even involuntary emissions in sleep were viewed with concern, and might be thought to be induced by evil female succubi stealing the man's vital powers. If they came from seeing a woman in a dream, the man should beware of

[18] R. H. van Gulik, *Sexual Life in Ancient China*, pp. 259, 353 ff.

meeting her in waking life for she might be a vampire or fox-spirit. Female masturbation, however, was viewed with tolerance, since the Yin supply of woman is unlimited. The texts mention dildos used for self-satisfaction by women, and pictures show them, but warning was given against excessive use which might damage 'the lining of the womb'.

Male homosexuality was not mentioned in sexual hand-books, since they were concerned with conjugal relations. From other sources it appears that it was rare in early times, flourished in the Middle Ages, and was not unusually frequent later. Homosexuality was tolerated among adults because intimate contact between two Yang elements could not result in loss of vital force. Some of the emperors had catamites, but also female concubines, and some T'ang poets seem to have been homosexual, though the affectionate language they used of their friends need not imply sexual intercourse. Male friendship was one of the Confucian virtues, and was often expressed in warm terms. A seventh-century story tells of an official and his wife spying on men bathing together, ostensibly to see if they had double ribs, but probably to observe how intimate they were together. Foreign observers in the nineteenth century said that pederasty was rampant in China, but social etiquette tolerated men walking hand in hand and homosexuals acted women's parts in theatres. There was probably a shortage of women in the ports where foreigners traded, and they would not see heterosexual relationships which were strictly private.

Female homosexuality was common and tolerated, and considered bound to prevail in women's quarters. Women could satisfy each other naturally or with artificial means, such as double dildos, or 'exertion bells' or 'tinkling balls' used for masturbation. The Chinese said that such artificial sexual aids were foreign: 'Burmese bells', 'Tartar pastures', 'barbarian soldiers', just as Europeans spoke of 'French letters' and 'lettres anglaises'.

Heterosexual intercourse, we see from the sexual manuals, was prepared and performed carefully, so as to activate the Yin essence. Kissing was important, along with further movement of lips and tongue in preliminary play. Foreigners thought that the Chinese did not kiss, because it was an act reserved for the bedchamber, and when the Chinese saw western women kissing in public they thought they were the lowest kind of prostitutes since even regular prostitutes would only kiss in private.

The sexual handbooks gave great detail of the various positions that could be adopted in sexual intercourse. Cunnilingus was approved, especially in Taoist texts, since it procured Yin essence for the man. But anal penetration and fellatio should only be used as preliminaries and if there was no complete male emission.

Single and married men were thought to be entitled to association with prostitutes, but it was different from conjugal intercourse as not designed for procreation and so it was not discussed in sexual manuals. Name-taboos did not apply, as the man did not know the prostitute's surname. Some writers thought that intercourse with a prostitute did not waste the male semen, since the woman from her profession had an abundant supply of Yin essence and gave back more than he lost to her patron. After the sixteenth century syphilis was identified and medical treatises warned of the dangers of association with prostitutes. Medical and sex books laid great stress on eugenics, in order to produce the best offspring.

Incest was rare, and to be punished according to the penal code as an 'inhuman crime' deserving death in a severe form, though some early imperial officials had incestuous relations with their sisters and other female relatives. Bestiality was also rare, though mentioned in connection with debauched rulers. Apart from sexual manuals, which were not pornographic, at least in intention, scatological material was rare and only found in some novels which delighted in exaggerated detail of male and female organs and secretions. The ancient Chinese had few inhibitions, though Confucian decorum sought to regulate natural functions.[19]

[19] On all this section see R. H. van Gulik, *Sexual Life in Ancient China*, pp. 47 f., 65, 160 ff.

REACTIONS

When he was preparing his masterly study of *Sexual Life in Ancient China*, R. H. van Gulik found 'that there was practically no serious literature available, either in standard Chinese sources or in Western books and treatises on China. The silence of Chinese reference books proved to be due to the excessive prudery that took hold of the Chinese during the Ch'ing or Manchu dynasty (1644–1912).'

The Manchus became masters of a divided China and transferred their northern capital to Peking, while negotiating with more stubborn resistance in the south. It was eventually agreed that intermarriage between Chinese and Manchus would be forbidden, a decree which remained in force till 1905. Chinese men adopted Manchu dress, but Chinese women did not change dress, and Manchu women were forbidden to wear Chinese dress or bind their feet.

Under foreign occupation the Confucian scholars insisted on the separation of the sexes, and everything belonging to sexual relations and women's affairs became taboo. The Manchus were persuaded to ban sexual manuals, and in due course they became even more adamant on this than the Chinese, though previously they had had few sexual inhibitions. Countless books gave information on all aspects of life under the Manchus, except sex. Sexual acts became especially secret, and this obscured their spiritual significance, and Chinese sexual life was regarded by outsiders as abnormal or depraved. European writers were quite misled by the puritanism and secretiveness about sex that obsessed the Chinese in recent centuries, and Chinese writers said this reserve had existed for over two thousand years.

Van Gulik and others managed to break through this reserve and study works that had long been hidden, and Joseph Needham revealed material on ancient Taoist sexual teachings and practices in his monumental series on *Science and Civilisation in China*. Fortunately for research workers banned Chinese texts on sex had been preserved in Japan, from as early as the seventh century A.D., and some Japanese and Chinese private collectors had erotic picture albums and sexological texts of later periods in their possession. Van

Gulik discovered that the ancient Chinese gave a great deal of attention to sexual matters, teaching householders how to conduct relations with women, and Needham held that Taoist teachings had influenced the development of sexual relations favourably and enhanced the position of women.

One result of the further subordination and seclusion of women was that love-making was often regarded as simply for male pleasure. Woman could be regarded as the enemy, and intercourse as the Battle of the Inner Chamber. Sexual practices became more and more explained from a male standpoint, and woman's role was to enhance the superior position of the man, in private as in public. In this century, under the various revolutions, the restriction of women has been formally abolished, along with concubinage and other practices which degrade women's status. But the strict modern sexual morality, which puts the state before the individual, owes something to both Confucian and Manchu morality, and the revival of Taoist joy in sex still seems to lie in the future.

Although for practised religion Taoism and its temples have been restricted or suppressed in communist China, much of the symbolism remains. In great national parades the flying ribbons and undulating banners recall the energies of Yin and Yang, balloons carrying slogans in bold characters are like esoteric cloud scripts, and pictures of leaders like gods are flanked with traditional Taoist signs for 'double happiness'. Communist ideology also, with its dialectics of nature, and the Hegelian principle of thesis, antithesis, and synthesis, may well prove adaptable to the Taoist theory of the harmony of Yin and Yang.[20]

[20] See R. C. Zaehner in *The Concise Encyclopaedia of Living Faiths*, 1959, pp. 410 ff.

Chapter 6

JAPAN'S FLOATING WORLD

SHINTO MYTH

The oldest Shinto mythologies, *Kojiki*, 'Records of Ancient Matters', and *Nihongi*, 'Chronicles of Japan', present names and stories of many heavenly beings in which male and female relationships and antagonisms appear. Out of the original formlessness the gods emerged, and two of them were the ancestors of all creation. These were Izanagi and Izanami, whose names have been translated as 'the Male who invites' and 'the Female who invites'. They stood on the Floating Bridge of Heaven, later painted and copied in art and temple bridges, and thrust down the Jewel-spear of Heaven into the ocean. Drops from the point of the spear, perhaps in phallic symbolism, became an island on to which these gods descended and where they erected a heavenly pillar and a palace.

Izanagi asked his companion, 'How is your body formed?' She replied, 'My body is well formed but it has one place where it is lacking.' Izanagi said, 'My body is also well formed but it has one place that is formed to excess. I would like to take my excessive part and insert it into the part where you are lacking and so give birth to the land. Let us walk round this heavenly pillar and have conjugal intercourse. You walk from the right and I will walk from the left and meet you.' They did so, but when they met Izanami spoke first, 'What a lovely youth', and Izanagi said, 'What a lovely maiden.' Then he protested, 'It is not proper for the woman to speak first.' They had sexual intercourse but produced a leech, which was not counted among their rightful children and it was sent away in a boat of reeds.

The two gods saw that their child was not good and reported it to the heavenly deities, who declared that it was because the woman had spoken first. So they came down to earth again and walked round the heavenly pillar as before, and when they met Izanagi said first, 'What a lovely maiden', and Izanami replied, 'What a lovely youth.' Another story says that these gods wished to have intercourse but did not know how to do it, until they observed a wagtail beating together its head and tail and by imitating its action they discovered the manner of coition and had many children. The wagtail was literally 'the love-knowing bird', sacred to these two deities, and there is a Wagtail Rock which is visited by pregnant women.

Commentators on the 'heavenly pillar' ceremony have said that it seems to have been an ancient practice to walk round a pillar before sexual intercourse, so that human practice was justified by divine example. Some agreed that the pillar was a phallic symbol, but the great scholar Motoori suggested that 'the man is above in sexual intercourse, like heaven or the roof which spreads over the house; the woman is below, like the supporting earth or the floor of the house. The pillar stands between them, strengthening and connecting top and bottom, and thus no doubt the idea is to strengthen and connect the couple.' Another scholar thought that this myth was a version of original sin, since Izanagi and Izanami were brother and sister and their union was incestuous, wherefore the circling of the heavenly pillar was a ritual designed to avoid this taboo. But both explanations seem forced and the phallicism would be more natural.[1]

Further myths were even more complex. Izanagi and Izanami gave birth to numerous deities, and she died after bearing the deity of fire. Izanagi went to visit his wife in the land of the dead, Yomi, lit a torch from his comb to see her in the dark, and broke a taboo by looking on her putrefying corpse. Then he fled, pursued by the hags of hell, but was saved by pelting them with peaches. Izanami herself chased

[1] *Kojiki*, tr. D. L. Philippi, 1968, p. 398; *Nihongi*, tr. W. G. Aston, 1972 edn. pp. 11 ff.

her husband and he pulled over a tremendous boulder to close the pass. The gods faced each other and broke their troth, with penalties of death and sexual pollution.

Izanagi then gave birth to many deities by bathing his body from pollution, including the sun goddess Ama-terasu and the storm god Susa-no-wo. Myths of the struggles of these two beings are even more popular than those of Izanagi and may originally have been independent, but they also reveal male-female tensions.

Susa-no-wo insulted Ama-terasu, breaking down her rice fields, strewing faeces in the hall of the firstfruits, and throwing a flayed pony into her sacred weaving hall. Ama-terasu was afraid and shut herself inside the heavenly cave. Then all the earth was dark, night reigned, the gods complained and all kinds of troubles arose. This struggle of light and dark, or sun and storm, is generally taken to explain eclipses, or storms, or summer and winter. Rituals were performed every winter in Japan to renew the sun's power, like winter fire ceremonies in northern Europe.

The eight million gods then met to perform magical rites to bring the sun back to the world. They made the birds cry, they took rock and iron from the mountains to make a mirror, they had strings of beads made, and they placed the mirror and beads in a tree with white and blue cloth. One god intoned a solemn liturgy, while a goddess jumped on a bucket and danced. She became divinely possessed, uncovered her breasts and pulled down her skirt to expose her genitals, a ritual perhaps of magic power to impart vitality to the deities.

The gods laughed loudly at this exhibitionism and the great sound aroused Ama-terasu's curiosity. She opened a crack in the cave door and seeing her beautiful reflection in the mirror she came to the door to get nearer. Then a god of great strength seized her hand and pulled the Sun Goddess out of the cave, while a rope was put behind her and a magical formula recited to prevent her going back. Now all the land became light and the gods imposed a fine on Susa-no-wo, cut off his beard and nails, had him exorcised and expelled.

This myth is of continuing importance, because Ama-terasu

105

is the supreme deity of the Shinto pantheon, and she was the ancestor of Jimmu the legendary first emperor of Japan. The principal shrine of Ama-terasu is at Isé, the most important of all Shinto shrines, set in spacious grounds in the south central part of the main island of Honshu. The sacred mirror kept there is the greatest divine symbol. Susa-no-wo has his shrine at Izumo on the northern coast, very ancient and greatly revered.

PHALLICISM AND DIVINE UNIONS

Phallic symbols were to be found widely in old Japan, as in China, India, Africa and many other lands, though nowadays they are often disguised or seen in abstract forms. In the ancient myth the Jewel-spear of Heaven may also have been the pillar round which Izanagi and Izanami walked before intercourse, or both spear and pillar may have been phallic symbols. The Japanese scholar Hirata suggested that the form of the Jewel-spear was like a 'male pillar' (*wo-bashira*), the name given to pillars or end-posts of bridges and staircases. This is a post which ends in a large knob, like a glans penis, and Hirata quoted a Chinese name for the penis as 'jewel-stalk'. These stylized male pillars are found everywhere still on bridges, railings, and balustrades though the symbolism may be unknown or forgotten. The *wo-bashira* was also the 'end-tooth' which Izanagi broke off his comb to make a torch to look at his wife in the land of the dead.[2]

Many of the Shinto deities had fertility functions, and the goddess who performed the strip-tease at Ama-terasu's cave did the same for a god of the crossroads. Phallic objects in stone and occasionally in clay were made in Japan from prehistoric times, and later such stones were sometimes fashioned into figures of old men to represent roadside guardians and fertility symbols for the fields. Especially phallic gods of Shinto were represented by phalli or human figures set up at the outskirts of villages to keep away

[2] See W. G. Aston, *Shinto*, 1905, pp. 186 ff; J. Herbert, *Shintô*, 1967, p. 150.

disease. Later there was some kinship or parallelism with the Buddhist Jizo, a bodhisattva of salvation, the Indian Kshitigarbha. Jizo was the protector of travellers, pregnant women, and children, and can also be seen in countless wayside stone figures as a monk. People put coloured bibs round the necks of these images, leave offerings beside them, and pile up stones for good luck. [3]

A class of gods known as 'preventive deities' were phallic. Often they had no temples but many images painted bright red or gilt, and festivals in their honour took place at crossroads. In olden days Road Festivals were held during epidemics, or before the arrival of foreign envoys, to ward against infection and demons. It is said that at the first full moon the boys in the imperial palace struck younger women with phallic-shaped sticks, to ensure fertility. Still at new year bamboo poles, with diagonal ends, are put outside houses and shops.

As in China the peach was a chief feminine symbol, like the apricot in India and the pomegranate in ancient Greece. Staves of peachwood were used for expelling evil spirits on the last day of the year. Rice could also be a female symbol, as it was formerly used in purification ceremonies and is still put in rooms where there is a newborn child. Beans were used in rites of cleansing and of expelling evil spirits. The comb, which appeared in myths, was another fertility symbol and had taboos.

Many writers testified to the presence of phallic emblems in Japan down the centuries. A book of 1795 wrote of carved wooden phalli eight feet long and four feet in circumference facing the road in Deha province. They were made afresh on the fifteenth day of the first month of the year. Slips of paper, of the kind used in many other Shinto shrines for good luck, were tied to such phalli secretly by women to gain handsome lovers. The great scholar W. G. Aston in 1871 found part of the road to Nikko, north-east of Tokyo, lined with groups of phalli, and a cave at the great Buddhist centre at Kamakura contained scores of carved stone phalli. Aston also witnessed a procession near Tokyo, with a phallus

[3] E. Kidder, *Ancient Japan*, 1977, pp. 40, 111.

several feet high and painted a bright red, the colour of male power, carried on a bier by a crowd of youths in festal costume, shouting and laughing, and zigzagging from one side of the road to the other. Similar ceremonies still take place in the country to encourage fertility, though their phallic symbolism may be denied by modern apologists.

From time to time there have been attempts to suppress the grosser manifestations of phallic worship. As early as A.D. 939 a large phallic image, which stood in a prominent place in Kyoto and was worshipped by travellers, was removed to a less conspicuous place. Since the revolution of 1868 Japanese governments have tried to repress notorious phallicism. Phallic images may be disguised by draping them with red cloth, or they may even be mistaken for the Buddhist Jizo. At Kamakura some temple courtyards still contain masses of small stone images draped in red, and though ostensibly they may be Buddhist deities they have a close resemblance to ancient phalli. Modern visitors to such temples may have no more idea of the phallicism than they have of its presence in the ubiquitous stylized bridge pillars, though perhaps the study of Freudian psychology may make people aware again of the ancient meanings.

There was strong phallicism also in Japanese folklore, in many stories of marriages between human beings and animals. There were fish, frog, stork, and fox wives, and also monkey, horse, and spider bridegrooms. But the commonest of all were the plainly phallic snake bridegrooms. One *Kojiki* story told of the god of Mount Miwa, who was notorious for his liking for beautiful women. He fell in love with a maiden and changed himself into a red-painted arrow, a fine phallic symbol. He floated down the river and struck her in the genitals while she was defecating. She took the arrow and placed it by her bed, whereupon it turned into a lovely young man who took the girl to wife.[4]

Another class of tale is of the so-called hemp-thread type. A girl was visited at night by a mysterious lover who got her with child. The parents wanted to discover his identity and

[4] *Kojiki*, tr. D. L. Philippi, 2, 53, and notes.

told the girl to sew hemp thread to his robe and follow wherever it led. She did so, and found that it led through the keyhole of the door to the shrine of the god of Mount Miwa, who had visited the girl in the form of a snake. This story and others like it are still told widely throughout Japan and neighbouring islands. It is clear that the god of Mount Miwa was regarded as a snake, or assuming the form of a snake. Today this mountain is the centre of a flourishing religious cult; it is infested with snakes who eat the offerings left by visitors.

In a variant of this story the girl followed the thread for miles till it disappeared inside a deep cave. The girl stood at the cave's mouth and called out that she wanted to see her lover's face, but a deep voice replied that if she did so she would burst with fright. However she persevered and at last a monstrous serpent crawled out of the cave, with a needle stuck in its throat from the thread. The girl fainted, but eventually she bore a child who grew up to be a huge boy and a great warrior.

It has been suggested that these stories were connected with rituals in which a woman was chosen and possessed by a deity associated with water and snakes. At other shrines there are often three deities, a mother, a child, and a god who is the child's father. Ritual possession and intercourse may lie behind what became popular tales; the women were 'possessed' in a double sense, spiritually by the deity and sexually by a priest. [5]

In other stories of divine marriage a goddess emerged from the sea to wed a man. There are many versions of this theme, and examples may be found in collections of folk tales, such as those by Lafcadio Hearn. The prototype of such a divine marriage was once again in the *Kojiki*. The daughter of the sea god left her water realm to give birth in a delivery hut, thatched with cormorant feathers, by the edge of the beach. She told her husband strictly not to watch her giving birth as she would revert to her original form. He thought this strange, watched her in secret, and was astonished to see her turn into a giant crocodile or

[5] C. Blacker, *The Catalpa Bow*, 1975, pp. 116 ff.

seamonster, and he ran away. When the goddess realized that she had been seen she felt extremely ashamed, left the child behind, closed the barrier between the land and the underwater regions, and went back into the sea. A modern Japanese scholar says that this divine-human marriage was designed to show that the imperial family, which came on its paternal side from the sun goddess, was maternally related to the sea goddess. The sea goddess was unable to subdue her yearning for her child, and she sent her sister to nurse it. The last word may be left with the forlorn husband who chanted,

> As long as I have life,
> I shall never forget
> My beloved, with whom I slept. [6]

IN AND YO

Shin-to (Shen-tao) was the Chinese name for the 'Way of the Gods' in Japan, to distinguish it from the 'Way of the Buddha'. It was not surprising that the idea of the Tao should have been introduced, along with other Chinese ideas in the flood of influences that followed the introduction of Buddhism into Japan. The *Kojiki* and *Nihongi* began with the primeval unity, which was followed by the separation of Heaven and Earth. Male and female then appeared, the In and Yo. Since these terms resemble the Chinese Yin and Yang it is natural to assume literary Chinese influence, though they were associated with the myths which are more native to Japan and also treat of male and female powers. Some modern Japanese scholars have rejected the Yin and Yang as foreign importations into native mythology, but as we have them the mythologies date from three centuries after the introduction of Chinese learning into Japan and there had been considerable mingling.

Public and popular life came to be influenced by the ideas of Yin and Yang. There was a department of it (Onyoryo) as early as A.D. 675, to advise the government on all matters of Yin-Yang lore, then the Taiho Code of 701–2 gave details for its organization, and exhorted diviners to 'be effective in

[6] *Kojiki*, 1, 45.

Yin-Yang divination, astronomy, medicine and fortune-telling.' In ordinary life the plan of a house and even the position of furniture was determined by Yin-Yang. For example, a chest containing valuables should not be placed in the southern part of the house, for that was the direction governed by the element fire and so liable to be burnt. The phenomena of the universe itself were explained in Yin-Yang terms, lucky days and years were chosen by such methods, and marriages were arranged according to signs related to them. Only in modern times, with the rejection of Chinese elements on the one hand, and the adoption of western ways on the other, has the Yin-Yang dualism been less regarded, though much probably remains in popular usage. [7]

A common symbol is a version of the Chinese *t'ai chi t'u*, the two pear-shaped halves of a circle which symbolize male and female, heaven and earth, Yang and Yin. In Japan this *tomoe* is often in three sections, sometimes associated with the Yin and Yang, and sometimes distinguished from it as having come from Korea instead of China. But perhaps the three sections are an elaboration of the original and more significant two. It is remarkable that even at the great shrine of Isé, said to have been maintained as pure Shinto unaffected by Buddhism, there are many of these symbols. They appear in decorations and dances at Isé, some of which it is admitted are of Buddhist and even of Indian origin. [8]

Like China, Japan was affected by a variety of Buddhist teachings and practices. Among those that affected sex, even marginally, the cult of tea which was introduced in the thirteenth century came to be regarded as an effective way of training young women in the etiquette of the hearth. Similarly, the strange connection between peaceful Buddhism and male warlike practices developed in the adoption of Zen Buddhism by Japanese warrior Samurai. Despite criticism from some Buddhists, Zen swordsmanship was justified as 'an art of protecting life' rather than a means of killing others, though it could lead to harshness both towards self and others.

[7] R. Tsunoda and others, eds., *Sources of Japanese Tradition*, 1958, pp. 59 ff.

[8] J. Herbert, *Shinto*, p. 150.

Although most Buddhist monks in Japan were fairly ascetic, there were sects which taught esoteric doctrines derived from China and India. In the eleventh century a priest Nin-kan founded the Tachikawa sect and taught a Japanese version of the left-hand Vajrayana with the help of a Yin-Yang teacher. They taught indulgence in the tabooed Five M's (see p. 37 above), and sexual intercourse as a means of 'directly obtaining Buddha-hood through one's living body'. Their doctrine declared that 'the Way of man and woman, Yang and Yin, is the secret art of becoming a Buddha in this life. No other way exists but this one of attaining Buddhahood and gaining the Way.'

Only a few of these texts survive; they are translations of Indian Tantric texts, made in China and imported into Japan by Buddhist pilgrims. They describe Tantric rituals and give illustrations of the 'sexual mandala', or 'the double mandala of the two worlds'. This shows a man and woman, naked but for ritual headdresses, lying in sexual embrace on an eight-petalled lotus flower. Although the man was on top, he had twisted round to place his head between the woman's feet and her head between his, while their outstretched arms and legs formed the lotus petals. The man's body was white or yellow, and the woman's crimson, while their meeting genitals were marked with the magical syllable *a*, which Tantrism held to be the beginning and end of all things. Other parts of their bodies were also marked with magical spells. A further picture of the 'Spark of Life' was a circle surrounded by flames, with a stylized sun and moon and two Sanskrit letters *a*, in white and red, which seem to represent semen and ovum, indicating the spiritual union of male and female principles based on biological union.[9]

This Tantric sect drew other elaborate correspondences of Yin and Yang, with the Mandalas of the 'thunderbolt' (*vajra*) and 'womb' (*garbha*) as symbols of male and female. The sect seems to have been popular for a time, but it aroused fierce opposition both from more orthodox Buddhists and from secular Japanese authorities. In the fourteenth century a notable Buddhist writer, Yukai, justified

[9] R. H. van Gulik, *Sexual Life in Ancient China*, pp. 358 f.

some esoteric Buddhism but attacked the Tantrism of the Tachikawa school. Nin-kan was said to have been guilty of 'some crime', and in exile he had made disciples of meat-eating, defiled people. Combining his notions with that of a Yin-Yang teacher from Tachikawa, the inner and outer learnings had been confused. 'They made outrageous assertions that the Buddha had previously taught their doctrines, a diabolical invention deserving of eternal punishment in hell.'

The Tachikawa sect seems to have practised sexual rites in mass-meetings, so that Japanese authorities banned the movement. An important Buddhist monastery submitted a memorial against the sect in the fourteenth century, and the leader was exiled and books of its teachings were burnt. It seems to have continued in secret, for even in the seventeenth century an orthodox Buddhist monk protested against Tachikawa practices, and traces of some of its doctrines still survive. Certain Japanese writers have said that the Tachikawa texts were invented by Nin-kan and his followers, but their correspondence to known Sanskrit sources shows that they were genuine translations. Only a small number of Tachikawa texts have been published, though they reveal some Chinese practices little known elsewhere, but there are still many unpublished texts in Japanese monasteries, sealed and labelled 'not to be opened'.

More orthodox Taoist sexual manuals were introduced to Japan, and often only survived in Japanese versions. In 984 a Japanese physician, Tamba Yasuyori, completed a compendium of medical science, *I-shin-po*, with extracts from several hundred Chinese works. This book circulated in manuscript for centuries, but a large blockprint was published in 1854. Sexual portions of this work consist of dialogues of the Yellow Emperor with the Plain, Dark, and Elected Girls. The Plain Girl said, 'All debility of man must be attributed to faulty exercise of the sexual act. Woman is superior to man in the same respect as water is superior to fire . . . Those who know the art of Yin and Yang can blend the five pleasures.' The Elected Girl said, 'The union of man and woman is like the mating of Heaven and Earth. It is because of their correct mating that Heaven and Earth last forever. Man, however, has lost this secret, therefore his age has gradually decreased.'

Then followed sections on nursing male and female potency, properties of their members, thirty positions, obtaining children, various diseases, and using drugs. There was the Chinese notion of *coitus reservatus*. The Plain Girl said, 'If a man engages once in the act without emitting semen, then his vital essence will be strong. If he does this twice, his hearing and vision will be acute. If thrice, all disease will disappear . . . If ten times, he will be like an Immortal.'[10]

WOMEN AND MEN

Although Japan has been very much a man's world, and still is in many respects, yet it has had outstanding women writers. *The Tale of Genji*, by the Lady Murasaki Shikibu, was written early in the tenth century A.D. It is a very long romance, with fifty-four chapters describing court life in Japan, and it is one of the earliest of the great novels of the world. While Japanese men were struggling to write in Chinese characters and reproduce Chinese ideas, women in aristocratic circles could write their native Japanese in newly-evolved phonetic script. Such women were relatively free, they were respected by nobles, and could become favourites at court or mothers of princes. Murasaki wrote in a living language, in complex sentences and with great sophistication.

Murasaki was born about 976 and died about 1015, but few details are known of her life. She married a distant kinsman in her early twenties, had a daughter, was widowed, and went to court in the service of the empress Akiko early in the eleventh century. Her novel told of the life and loves of Prince Genji, of which there were countless charming sketches and details, with assignations and nocturnal visits. Sex was agreeable and little restricted, but religion only came in incidentally, and there was no Yin and Yang.

The second chapter of the *Genji*, 'the Broom Tree',

---

[10] R. H. van Gulik, *Sexual Life in Ancient China*, pp. 122, 135 ff.

presented the prince on a rainy night discussing with a friend the different categories of women. There were the young and pretty, soft and feminine, gentle and childlike, chilly and unfeeling, quiet and steady. One example was given of a man who was really kind and sympathetic, but he committed 'some minor dereliction', and his wife ran off to become a Buddhist nun. Her hair was cropped but she could not hold back her tears and regretted the life she had left behind. 'The Buddha is not likely to think her one who has cleansed her heart of passion.' But her husband had not lost his affection for her, he sought her out and brought her back, because 'the bond between husband and wife is a strong one.' It would be as foolish for such a woman to let his 'little dalliance' upset her, as it would be for him to allow the memory of her flight still rankle. So the wife should be docile and trusting, and the man with tactful guidance would mend his ways. The lady writer importantly gave the views of both sexes.

From the same period came another work in pure Japanese, the vivacious *Pillow Book* or 'Miscellany' of Sei Shonagon. This depicted court life, with little hint of the world outside. It was full of sketches of society and nature, and although it tended to scorn the lower orders it was also critical of men. Men had strange notions, they ate in disagreeable ways, they left attractive women for ugly ones, they might be either affected or fail in etiquette. A preacher, however, ought to be good-looking, for otherwise people might not keep their eyes on him. But priests on night duty might be confronted with shameful things, overhearing young women joking about other people. It was also very annoying when visiting a temple to gaze on the sacred countenance of the Buddha, to find a throng of commoners in front of the image, incessantly standing up and prostrating themselves.

More continually autobiographical were the *Confessions* of Lady Nijo at the end of the thirteenth century. She told how she became a concubine of a retired emperor in Kyoto at the age of fourteen and ended, after several love affairs, as a wandering Buddhist nun. She began with the emperor coming into her young bed at night to her great surprise. He did not force her then, but returned the next night and 'he treated me so mercilessly that my thin gowns were badly ripped.' She wept, but stayed at court and exchanged poems with the emperor. She became a court lady and was independent enough to take lovers, but finally when the emperor's ardour had cooled, and owing to the jealousy of the empress, she was forced to leave the palace at the age of twenty-six. In the last nine years of her short life Lady Nijo became a nun and went on pilgrimage to both Buddhist and Shinto shrines. She met the emperor again after he had taken religious vows, visited him on his death bed, watched his funeral from the outside, and attended services marking the third anniversary of his death.

These and similar writings provided vivid pictures of court life and romantic affairs, but they told little about other classes. The Samurai warrior class joined Japanese feudal traditions with Confucian ethics in the service of the emperor. This ideal was expounded in the seventeenth century by Yamaga Soko in what later came to be known as the Way of the Warrior (*bushi-do*). A strong sense of duty put loyalty ahead of personal gain, and a readiness to meet death at any time. Yamaga taught morality and martial discipline, but he also stressed the peaceful arts as important to the Samurai, and so typified the change of the Samurai from a purely military aristocracy to intellectual and political leadership.[11]

Samurai women, however, probably had less freedom than those in the noble classes above them or the great mass of merchants and peasants below them. Their life was one of dependent service upon their husbands, following their ideals of honour. When girls reached womanhood they were given pocket daggers to defend their chastity against assailants. They learnt, like their husbands, how to commit suicide, where to cut their own throats, and how to tie their lower limbs together so that the corpse would be found with the limbs properly composed. Stories were told of women who overheard plots to assassinate their husbands and in the dark they substituted themselves to die in their place.

The position of women was often seen as depressed. The

---

[11] R. Tsunoda and others, eds., *Sources of Japanese Tradition*, pp. 394 ff.

term 'woman' was applied to a slow and stupid man, and the word 'noisy' repeated three times the Chinese character which represented woman. From adolescence a wall was raised between boys and girls, and even in marriage a woman had little formal status in the home. Neo-Confucianism may be blamed for some of the male dominance, but much seems to have been native Japanese, and due both to the greater physical strength of the male and the traditions of feudalism. Even today Japanese women often complain that they have no voice in the education of their children and their choice of partners, though strong-minded women perhaps do make themselves heard.

Buddhism in Japan affected women in different ways. The founders of some great monasteries decreed that they should never be visited by women for they were regarded as a source of defilement, probably because of menstruation. Women were excluded from some important offices and religious observances, as in some other religions. However, the spread of Pure Land or Amidist Buddhism gave women equal rights to salvation along with men, and it found new leaders not in monasteries but living in society much like lay people.

Also since ancient times there were very important women in Japan, 'female shamans', who were 'possessed' by spirits and exerted great influence. The Shinto *miko* was a powerful sacral medium who served in shrines and acted as a mouthpiece for the gods or ancestors. Her prototype may be seen in a description of the goddess Ama-terasu in the *Kojiki*:

> Undoing her hair she wrapped it in bunches, and in these bunches, on the vine securing her hair, as well as on her left and right arms, she wrapped long strings of beads . . . Shaking the upper tip of her bow, stamping her legs up to her very thighs into the hard earth, and kicking it about as if it were light snow, she shouted with an awesome fury. [12]

The *miko* was to be found in the imperial court, giving the emperor instructions from the gods, as well as in countless

[12] *Kojiki*, 1, 14; see C. Blacker, *The Catalpa Bow*, pp. 104 ff., 130 ff.

towns and villages where she acted as intermediary between the local gods and the villagers under their care. Many *miko* wandered about like strolling minstrels, to deliver spiritual messages and they were regarded as oracles.

The activities of the *miko* were supposedly suppressed in 1873, in an official attempt towards the 'enlightenment' of the country and the purgation of Shinto from both super-stition and Buddhism. However, the so-called 'new religions' which have flourished in Japan since then have been especially inspired by similarly possessed women, who to varying degrees continue something of the old traditions. The largest and most prosperous of these, Tenrikyo, was founded by a possessed woman, Nakayama Miki. After youthful sufferings, in 1838 she fell into a violent state of trance, and from that time she had further fits of possession, she developed healing powers, and in 1869 she began to write automatically a long poem, *Ofudesaki*, of 1,711 verses, which became the scripture of her church. Sex is no barrier, therefore, in these religions, but on the contrary it is often a great help and inspiration.

MARRIAGE

Shinto has emphasized life-giving ceremonies, with blessings at the birth of babies and at regular intervals. For centuries the Japanese have taken their children to the shrines at the ages of three, five, and seven, on the fifteenth day of the eleventh month, to thank the gods for health and pray for continuing protection. At the New Year families take several days holiday to visit the shrines and ancestral homes, where food and drink are taken and arrows are bought at the shrines as virile symbols of the future. Japanese Buddhism, on the other hand, has appropriated practically all the funerals, in a negative complement to life-seeking Shinto.

It is surprising, therefore, that marriage ceremonies do not seem to have had a prominent place in traditional Shinto shrines, and W. G. Aston went so far as to say that 'Shinto never had a marriage ceremony. No Shinto or other priest is present.' Yet nowadays weddings are held at Shinto shrines, and I witnessed a splendid ceremony at the great Atsuta shrines in Nagoya, where the sacred national sword is preserved. This was a very expensive occasion, with the bride in gorgeous white and red costume, while the bridegroom, as in some other countries, wore much duller robes of grey and black. There was a great reception, with many presents, and the cost of such an event may be one reason why some Japanese now go to a Christian church for a 'white wedding', whicn is formal but cheaper.

As in many cultures, Japanese marriage was primarily an arrangement between two families, and was traditionally arranged by go-betweens. The wedding was therefore a social contract, with presentation of gifts, special dress, banquet, and drinks. After the bride in all her finery is fetched in a litter or bus to the groom's house they sit in front of the family *tokonoma* alcove, which is decorated with lucky pine, bamboo, and plum, and hung with pictures or scrolls. Here, or behind a screen, is performed the marriage ceremony of the 'three-three-nine times' drink under the direction of the go-between. Two girls pour out *saké* wine into a tier of three cups, one on top of the other. The groom and bride drink this in turn, indicating that they will share joy and sorrow, and after the third cup the go-between announces that they are properly married. At the end of the banquet young men may bring in the nearest stone Jizo, the Buddhist children's god or former phallic symbol, on which obscene verses are pasted. Some days later the Jizo is put back, with a new bib sewn by the bride.

In former times the marriage was consummated in a special nuptial hut, perhaps because of the fear of contamination by sexual energy. The storm god Susa-no-wo was said to have made a house with a manifold fence when he married a goddess, and these two are now sometimes thought to preside over marriages. In ancient Japan menstruation and childbirth were regarded as polluting and both menstruating and pregnant women were required to live in a hut apart from the main building and eat their food separately from others.

Boys and girls traditionally had been allowed to mingle until five or six years of age, but after that the girls withdrew to their own company and from ten years on were forbidden to play with the opposite sex. It was less restrictive in the country where young men and women worked together, and in modern times they mingle at least on the way to school and in social occasions. Schoolgirls seem to speak much more freely to visitors, asking questions, standing with them, and having group photographs taken, while the boys listen silently to preserve their status. Clandestine affairs between young people were probably easier in the past before electricity made them difficult to hide, since the light is often left on all night. Love letters are written, and denied, and secret affairs may not get beyond this stage. If a girl becomes pregnant her parents try to arrange a quick marriage, or she may be sent away to become a concubine or geisha, after the birth of the baby. Trial marriages may take place, with only a small exchange of gifts and no special clothes or hairdressing.

The strains of modern life are illustrated in novels, such as *The Makioka Sisters*, where the obstacle was the second sister who had to be married before the third could wed. Many attempts were made by go-betweens, prospective bridegrooms inspected, and their background investigated by detectives, but the sister remained shy and difficult, getting older until she had to accept an inferior husband to one she might have had. Meanwhile, in the stresses of life in Osaka city, the younger sister had an affair and a child, and the tensions of old and new and east and west are skilfully drawn.

A slight religious element has been added to wedding ceremonies in placing on a stand or table a scroll picture or doll figures of the old man and old woman of Takasago, spirits of two ancient fir-trees, who are the Darby and Joan of Japanese legend. Their story is chanted at weddings and anniversary parties, it is related in one of the most beloved No plays, and it was painted by the great Hokusai and later artists. It is the tale of the imperishable love and calm happiness of two old people, the old man (*okina*) and the old woman (*uba*). He carries a rake and his wife a broom, they stand in front of two entwined pine trees, with cranes and turtles in the sky and on the ground. The pine tree was thought in China and Japan to be full of magical power because it does not wither, and tortoises and cranes also had a large amount of such power because they lived so long. A setting red sun in this picture brings on the evening of the happy faithful life together. It is a gracious ideal, which offsets the contrary facts of concubines and geishas.

FLOATING WORLD AND GEISHAS

In the late seventeenth century the term *ukiyo*, 'floating world', was used in Japan of the pleasant but changeable state of society. It had been given earlier to the 'sorrowful world' of Buddhist descriptions of dust and grief. The new word arose from a pun between sorrowful and floating, and it depicted the unstable society which had succeeded the medieval world. *Ukiyo* was used of brothels and places of licensed amusement which were prominent in urban society. The word was also applied to many products of culture, including *ukiyo-e*, the woodblock prints which are the most famous artistic products of the period.

Outstanding examples of *ukiyo* or demi-monde fiction are found in the writings of Ihara Saikaku, who portrayed the two great interests of the floating world, sex and money. His first novel was *The Man who spent his Life at Love-making*, telling of a hero who roamed around the country making love to thousands of women and hundreds of boys. The book has been compared to *The Tale of Genji*, but whereas Genji's women were distinguished by their family, accomplishments, and tastes, Saikaku entered into precise details of the beauty of their bodies.

In his more realistic novel *Five Women who Loved Love*, Saikaku was both poetic and popular. Instead of the court of Genji, there were teahouses, bathhouses, theatres, brothels, and the homes of commoners. Such life came to be known as Chonin-do, the 'Way of the Townspeople', in contrast to Bushi-do, the 'Way of the Warrior'. Although not devoutly religious, the preoccupation with sex here had an intensity akin to religious feeling, and there was a connection between the Buddhist 'sorrowful world' and the 'floating world' of fashion in that both were ephemeral, and Saikaku for all his eroticism felt the vanity and pathos of life.

121

In *The Mirror of Manly Love*, Saikaku treated of homosexuality. The growth of Buddhist monasticism had made it common, between master and disciple either latent or overt, and in the warrior class young men gave their service in exchange for the protection of elders. Saikaku wrote of Buddhist temples and Shinto shrines as favourite haunts of homosexuals, but the theatres did even more to promote the love of men for men and make it appear less unnatural than in society as a whole. To this day the No and Kabuki theatres ban women actors and employ the *onnagata*, 'female form', the male actor of female parts. Handsome young men adopt the dress and mannerisms of women and often act such parts even off the stage.

Religion seemed to be incidental to this sensual world, and yet it served rather as a basis and background. For the Buddhist belief in *karma*, the chain of moral causation, permeated many of the stories. Life is conditioned by actions in a previous existence, and morality here will determine the next life. This might be an intolerable burden, worse than Original Sin, were it not for the appearance of Saviours, especially Amida Buddha who applied his own great merits to the salvation of all beings so that they might share the joys of his Pure Land. Before that happy state arrived, ruined lovers could resort to Buddhist monasteries and there pray for reunion with their beloved in paradise.

Ukiyo books consisted of light-hearted and sometimes pornographic material, and similar though seemingly opposed works were those which had the avowed intention of 'encouraging virtue and chastising vice'. Both Buddhist and Confucian precepts were used to emphasize duty and control of the passions, though they could suggest the pleasures they condemned. Books such as *The Ukiyo Bathhouse* or *The Ukiyo Barbershop* tried painlessly to inculcate moral ideals, showing their heroes tempted in many ways, but following the Buddhist conviction that evil leads to evil and only good produces good.

Erotic paintings had long been current in Japan, with Chinese influence but also with distinctive Japanese traits. An early erotic scroll is of the thirteenth century but said to be a copy of a Chinese original of the tenth century. This was the *Scroll of the Initiation*, giving sixteen pictures of sexual intercourse in different positions performed by a courtier and lady. The positions can be paralleled in Chinese and Indian books, but the Japanese scroll shows enlarged sexual organs such as were characteristic of early and later Japanese erotic paintings. A Chinese novel of the sixteenth century, *The Joys of Man*, survives in a Japanese adaptation accompanied by small erotic pictures, and without giving obvious Taoist teaching it stressed the therapeutic properties of the sex act and the importance of conserving the semen. It is curious that some of these erotic novels were printed in Buddhist temples, as late as the nineteenth century, and with the same antique movable type that was used for printing Buddhist holy books.

In old Japan, as in medieval China, women were often painted as sturdy, with chubby faces, full breasts, slender waists but heavy hips, suggesting child-bearers. Men were likewise shown as virile or martial, with thick beards and strong bodies. But by the seventeenth century the Chinese ideals of female and male beauty changed to the other extreme. The Japanese followed this fashion, and the later *ukiyo-e* prints depicted it, with frail-looking women whose oval faces were considered the height of beauty.

Among many *ukiyo-e* artists, including some great painters of nature, Utamaro in the late eighteenth century more than any other revealed the life of women in busy commercial Tokyo, particularly those of the Yoshiwara, the regulated brothel quarter, and neighbouring districts. He showed great courtesans, but also daughters of wealthy merchant families, favourites of unlicensed teahouses and prostitutes of lower grades. Utamaro concentrated on women, not merely to provide the public with pictures of pretty girls but to study individuals in countless activities. His work ranged over the changing seasons and annual rituals: women playing shuttlecock, viewing cherry blossoms, sitting by the river, looking at the moon, catching fireflies, drinking *sake* in the snow, going to temples, attending religious ceremonies, celebrating the five traditional festivals of the year, making trips to scenic spots, going to the theatre, and visiting the gay quarters. Women were shown in the home also: cooking, doing needlework, spring cleaning, or rearing silkworms. They were painted in different poses: plucking their eyebrows, making up their faces, resting under mosquito nets, waking in the morning, washing their hands, and doing manual work. Utamaro's 'spring pictures' (*shunga*) became famous and his series of twelve colour prints known as 'The Poem-Pillow Picture Book' was perhaps the greatest of all erotic *ukiyo-e*.

Utamaro painted entertainers and prostitutes, and nude women divers. These works showed bare-breasted teahouse girls, no-nonsense smokers, sensual wantons, and tired girls going to bed. Through the superficial brilliance there were often hints of personal sadness or tragedy. The women were restricted in many ways, cut off from their families, known only by fanciful names, and forced to give themselves to a succession of men. They existed at a lower level than men, looked down upon as unspiritual and considered by orthodox religion as sinful and defiled. Utamaro showed great sympathy and did not treat women as objects, but he demonstrated clearly that other men did so and that women were their playthings.

This floating world was a reaction from the formality of home life. Japanese women were married for family reasons, and it was impressed upon them that their task was to wait on their husbands and bear them children. Long education made them repressed and submissive, and the men's home lives were restricted by tradition, so that they looked for entertaining female company, with elegance and humour that were missing in the wife. The gay quarters provided an escape from reality, at the expense of the women there and at home. Only the upper classes could afford mistresses and concubines, but middle-class men sought outlets with geishas or prostitutes.

A geisha was an 'art person', a dancing girl, and strictly a professional dancer and singer. The word was often loosely used, however, either of a high-class courtesan or a low-class prostitute. A visit to a geisha house gave entertainment, but not the right to automatic sexual intercourse. For that, a man would have to sign a contract in which the geisha would become his mistress for a time. At geisha houses nevertheless the dances, songs, gestures, and repartee were traditionally suggestive, expressing things that a wife would not say, giving relief from the 'circle of duty' into the 'circle of human feelings'. After an evening at a geisha house some men might return home, and expect their wives to wait on them, others might visit prostitutes. Yet others would be faithful to their homes and would never visit geishas.

Geishas have their own special gods, particularly the rice god Inari, symbolized by the fox, who is enshrined in geisha houses and brothels. They go to the temples in gay kimonos on the first and fifteenth of every month, to pay their respects. After praying, they sit on the platform talking and smoking before walking back home. Geishas are rarely married, and if poor people sell their daughters as geishas to some distant town the girls rarely return home.

In Japanese peasant life the role of women has often been more free than in town or higher society, and there might be close partnership between husband and wife. A peasant woman might laugh and joke as freely as any geisha, and use bawdy language that would horrify a well-bred woman. At wedding parties peasant women might make broad jokes at the expense of bride and groom, drinking plenty of *sake*, and while men might sing or act some harmless dance married women might dance in imitation and exaggeration of copulation, to the accompaniment of a loud female chorus and roars of laughter.[13]

For the scores of millions who live in towns the 'new religions' provide society and release from pent-up feelings. Most of these movements have huge temples, built by voluntary labour and gifts, and including social and educational centres. The lavish buildings contrast with the tiny homes of their worshippers, but they provide luxury and peace in communal life that is hard to find elsewhere. People visit not only the worship halls, but the adjoining rooms and balconies which are carpeted so that everyone can take off shoes as at home, women can put down their children, and all can rest, play, eat, and learn in peace. There are often counselling sessions, open every day, in which men and

[13] J. F. Embree, *A Japanese Village*, 1946, pp. 155 f.

125

women receive guidance in personal problems. One of the largest Buddhist societies in Japan uses the symbolism of the 'circle of harmony' to depict both the 'wheel of the Law' and the groups of its counselling sessions. [14]

[14] K. J. Dale, *Circle of Harmony*, 1975, pp. 37 ff.

Chapter 7

TRADITIONAL AFRICA

ATTITUDES

Edwin Smith, in a classic and exhaustive study, declared that

> to write of the Ba-ila and omit all reference to sex would
> be like writing of the sky and leaving out the sun; for sex is
> the most pervasive element of their life. It is the at-
> mosphere into which the children are brought. Their early
> years are largely a preparation for the sexual function;
> during the years of maturity it is their most ardent pursuit,
> and old age is spent in vain and disappointing endeavours
> to continue it . . . To them, the union of the sexes is on
> the same plane as eating and drinking, to be indulged in
> without stint on every possible occasion.[1]

That was over sixty years ago and many things have
changed. The Northern Rhodesia in the title of Smith's book
has become Zambia. Many of the Ba-ila are now Christians,
have been affected by a puritan ethic, and might deny that
the above was a correct picture either of the past or the
present. Smith himself was a missionary at first, though he
left his society and became more of an anthropologist. Yet he
wrote that 'there is much that is unpleasant in this part of
our subject', and he put some of the descriptions of sexual
detail into Latin, as did van Gulik in his study of sexual life
in ancient China.

Africa is important because it probably contains larger

[1] E. W. Smith and A. M. Dale, *The Ila-speaking Peoples of Northern
Rhodesia*, 1920, ii, p. 35.

127

numbers of illiterate tribal people, still wholly or partly outside the scope of literate historical religions, than other continents. Australia has now relatively few aboriginals, and while larger groups of tribal peoples are scattered about Asia and America, there are perhaps over fifty million such peoples in Africa. There are up to a hundred million Christians in Africa, and more Muslims, yet ancient beliefs and customs still affect many of these also. Yet a major difficulty of studying African ideas and practices is the absence of texts from the past, for since the art of writing had hardly penetrated the tropical and southern parts of the continent before modern times, there are no old sex manuals, books of teaching, or classic novels, to illustrate sexual ideas from inside Africa. One source of knowledge is in art, wood carving and stone sculpture, though the interpretation is not always easy; but its contribution to understanding sexual attitudes is helpful.

The great diversity of African peoples is a further handicap, since there was no overall organization and it is not possible to give more than a few examples from widely separated peoples. This obtains to some extent in India, and even in China, though the latter had much more cohesive imperial rule. In Africa, anthropologists have made many studies of modern customs, and some of these may go back to ancient roots. But even anthropologists do not always discuss sex in detail, or not in relation to religion, and some of the most sympathetic tend to be reserved on the subject. Smith's book was a landmark, and he had the great advantage of having been brought up among the Ila-speaking peoples and knowing their language as intimately as his own. And lest it be thought that he suggested unlimited sex in the quotation given above, it must be said that he recognized that as there are limits to eating and drinking, so there have been to the indulgence of sexual instincts. The proprietary rights of others had to be respected, and there were taboos on the times and places of sexual intercourse.

MYTHS

Among the Ila-speaking peoples the name of the supreme

deity was Leza, and variations on this name have appeared widely in eastern and southern Africa. Although various derivations of this name have been suggested, Smith favoured its origin from a verb meaning 'to cherish', as a mother does her favourite child or a chief his community. Leza lived in heaven, was the creator of all things, and sent rain and help to the earth. Although Leza was generally regarded as male, a father, and distant from man and his ways, there were myths which spoke of his family, wives, and sons. In one story at least Leza appeared as 'the mother of all beasts'.

Leza established human customs, sent death, and also provided man with medicine to promote birth and propagate the race. But it is remarkable, if Africans have been obsessed with sex, that there seem to have been few myths, here or elsewhere, of sexual adventures of the supreme God. There is no known parallel in Africa to the amorous exploits of great Indian deities like Shiva and Krishna. True, there is sexual energy and phallicism of some lesser African divinities, but little of the supreme God.

In many parts of Africa God was regarded as male and father, but in some places the deity was female. In the Niger delta the chief divinity was a mother with many breasts and children, and among the Igbo the earth goddess, Ala or Ale, is the Supreme Being. In neighbouring Dahomey (recently renamed Benin) the chief of the gods are the pair Mawu–Lisa (no relation to Leza), who are parents of other divinities who also went in twos. There is some reference in myths to an older androgynous deity which gave birth to the dual creator and then disappeared. Mawu–Lisa would then be the arrangers of the world from pre-existing material, rather than creators *ex nihilo*. Mawu was the female principle, with fertility, motherhood, and gentleness, like the moon. Lisa was power, warlike and tough, like the sun. Together they assure the rhythm of day and night and, by presenting their two natures alternately to men and women, the divine pair express complementary elements in life.

The dual nature of the Dahomean gods was seen as reproduced in mankind. The ideal birth was a twin birth, and the cult of twins was widespread. In other parts of Africa twins might be thought to be dangerous and they were exposed. Further, the universe in Dahomean and some other African cosmology was compared with the two halves of a calabash, one on top of the other and meeting at the horizon. The earth floated within this calabash, and round it was coiled a life-force personalized as a snake. This serpent also had two aspects, male and female, thought of as twins, or perhaps having a dual nature rather than separate elements. The rainbow was an expression of this snake, the red part being male and the blue female.[2]

Among the Dogon of the western Sudan there was also twin-ness, and more exceptional sexuality in cosmology. The supreme God, Amma, created the earth from clay and it lay flat like a female body, face upwards. Its vulva was an ant-hill and its clitoris a termite-hill. Amma was lonely and approached the earth to have intercourse, but this was a mistake. As Amma drew near the earth's termite-hill rose up, and being as strong as his penis intercourse could not take place. Amma was omnipotent and cut down the clitoris so that the union could proceed. The clitoridectomy of women is justified from this divine example. But from the divine blunder there was born, instead of the intended twins, a jackal which brought constant trouble to God. After further intercourse, with the former barrier removed, normal conception took place. The divine seed, water, entered the womb of the earth and twins were born. Like all other creatures, these twin beings were images of the twin-ness in creation and each had two spiritual principles of the two sexes.

The Dogon are said to believe that man, like the primordial beings, has two souls of opposite sexes. One lives in his body and the other lives in the sky or water. When a boy is circumcised he is freed from the element of femininity, which he had had in childhood. Similarly when a girl is circumcised or excised she is freed from the male element, and her clitoris no longer prevents intercourse. At circumcision prayers are offered for the stabilization of the

---

[2] P. Mercier, in *African Worlds*, ed., D. Forde, 1954, pp. 219 f.

soul, of the boy or girl, and spiritual force is thought to be released.

This Dogon mythology has been written up in great detail, and some of its apparently romantic elements have been criticized by other anthropologists, as having little parallel elsewhere in Africa, but it does perhaps show the relationship of myth to ritual, or derives myth from ritual. It is not possible to give other examples here from this vast continent, for reasons of space and material.[3]

PHALLICISM

Among the Ashanti of Ghana a myth was told to explain the origins of sex and family. It was said that long ago a man and a woman came down from heaven, and another man and woman came up out of the earth. Such dual origin of mankind appears in other traditions. Later, the supreme God sent down a python which made its home in a river. At first the men and women lived together without intercourse and they had no knowledge of conception and birth. One day the python asked them if they had any children, and being told they had not he said he would show them how to conceive. He made the couples stand face to face, then sprayed water on their bellies with ritual words, and told them to return home and lie together. From this phallic example, they learned intercourse and brought forth the first children.[4]

The python, a non-poisonous snake, is sacred to these clans, it is never killed and if a dead python is found the people sprinkle white clay on it and bury it. The cult of the python is widespread in western Africa and there is a famous snake temple at Ouidah (Whydah). This was described as long ago as 1705 by the Dutch trader William Bosman, who said that it stood under a very beautiful lofty tree, as it still does, and contained a sort of grandfather of all snakes, 'as thick as a man and of an unmeasurable length'. Bosman claimed that the King's daughter was captured for

[3] M. Griaule, *Conversations with Ogotemmêli*, 1965, pp. 17 ff.
[4] R. S. Rattray, *Ashanti*, 1923, p. 48.

the snake and finally displayed with all the other girls, 'naked, except only a silk scarf, which was passed between her legs, and richly adorned with coral'.

When I visited this temple over thirty years ago it was in a small mud-walled compound opposite the Roman Catholic cathedral. The priest willingly fetched out pythons to show to visitors; they wander tamely about the town and are taboo as food. If a man meets a python he bows down before it and says, 'My father'. The python is called 'snake of the forest' (*dangbe*), and its devotees are 'snake-wives' (*dangbe-si*), but they are male and female and there is no suggestion of a sexual cult. Dedicated followers of other gods are also called 'god-wives'.

It has been seen that another snake appears in Dahomean mythology and may have some phallic significance. The great snake Dan coils round the earth to give it life and stability; there are said to be 3,500 coils above and a similar number below. In another form of the myth Dan set up four pillars at the cardinal points which support the sky, and twisted round them threads in the three primary colours, black, white, and red, which are the colours of the clothes that Dan puts on at different times: night, day, and twilight. Dan accompanied God in creation, carrying him from end to end of the earth, and wherever they stopped a mountain arose out of Dan's excrement, which is the name also given to metals and precious stones found in the earth.

Cylindrical objects of wood and stone are to be found in many parts of Africa, though it is not always sure that they are phallic, since there are many human figures also and the human form may or may not be regarded as phallic. Images of human figures often show natural or enlarged genitals, according to the force that the artist wished to suggest. Stone figures in Guinea and Sierra Leone which have been called phalli, basically of cylinders surmounted by featureless heads, are probably simple human figures and where more detail appears the genitals are rarely shown. Many famous stone, bronze, and terra-cotta figures of Ifé and Esié in Nigeria mostly represent human beings, but there have been some clear stone phalli, notably in the market at Ifé. It is generally agreed that stone pillars among the Ekoi of the

132

Cross River are basically phallic, with the glans sometimes modelled into a human head and a protruding navel added.

Stone columns with rounded tops have been found widely in Africa, from Ethiopia to the western Sudan. From the ruins of ancient Zimbabwe in eastern Africa there have been recovered many small conical stone objects which have been called phalli, though some scholars consider that they are highly stylized female figures. One informant recognized these as similar to objects made in clay and given by mothers to their daughters as playthings, though eventually they were used in ceremonies of initiation into womanhood. [5]

Many clay, wooden, stone, and ivory carvings of human figures have been made in Africa. They are often nude and have naturalistic details of genitals, as in popular twin (*ibeji*) images. Such twin figures were placed outside the door of the home of the twins and were given regular offerings to maintain their health. If one twin died, the mother would carry the figure of the twin in her skirt belt. Some carvings showed men and women copulating, as on the carved wooden doors of palaces at Ifé. Marcel Griaule's informant explained the symbolism of a Dogon village in terms of human bodies, with the stone for oil-crushing as female genitals and the village altar as a phallus. [6]

The most obvious, and often exaggerated, phallicism may be found in many wayside shrines in West Africa. Along the coast of Ghana many images outside towns had phallic appearance, with clay figures that had carefully-shaped wooden phalli of exaggerated size, with bundles of similarly shaped wooden rods laid in front of them. Nowadays, such rods are placed there to copy ancient styles and it is said that they are simply cudgels to hit an enemy. In many parts of Dahomey and Nigeria there are countless images of a spirit called Eshu or Legba, which commonly have a prominent phallic element. This is a guardian spirit, dangerous to strangers but propitiated by householders and villagers. The basic form is a conical mud pillar, bearing rounded marks or cowrie shells, and if it is in the street it is protected by a small

[5] P. Allison, *African Stone Sculpture*, 1968, pp. 52 ff.
[6] M. Griaule, *Conversations with Ogotemmêli*, p. 95.

thatch or tin roof. A household Legba is always in human form, generally sitting with hands on knees and showing a large wooden phallus. One I saw on a main road in Porto Novo was a life-sized red clay figure of a European, wearing a sun helmet and wristwatch, and with a big phallus.

Early missionaries were shocked by such images and Stephen Farrow, writing of his experiences in Nigeria in the 1890s, denied that Eshu was a phallic god and declared, 'He is really the deity of supreme wickedness' and 'the prince of darkness'. Such diabolical and dualistic ideas were foreign to the Nigerians at that time, though a later tendency has been to play down the phallic element and E. B. Idowu called Eshu simply a 'special relations officer', running errands between heaven and earth. The silence of today is as significant as the shock of yesterday.

INITIATION

Connections of religion and sex appear in initiation ceremonies, when young men and women are trained in the mysteries and relationships of adult life. Ba-ila boys were taken to the cattle outpost and spent some time there in trials of manhood. They were not circumcised, like some of their neighbours, but had to run a gauntlet from older men who beat them with sticks and pelted them with stones. After sleeping on the ground they had to bathe in cold water, and for several days they lived together naked day and night. They were given medicine for the first emission of semen, to rub on the scrotum and to drink. Pubic hair was plucked out with the idea of preserving strength, and this practice was continued by both men and women to maintain cleanliness. Instruction was given by elders in sexual and moral matters, by an initiator who was called a 'husband', and who imparted traditions as the words of God.

In southern Africa the Xhosa and Basotho still practise male circumcision, whereas some other peoples never had this practice or have been persuaded by missionaries to abandon it. Circumcision was regarded by the Xhosa and Basotho as the formal incorporation of young men into tribal life, without which they could not be married or inherit family property. An uncircumcised man would be regarded as still a boy, or even called 'a dog' or 'an unclean thing'. A woman would not marry an uncircumcised man, since her family would not negotiate dowry with such a person. Even in modern times men who have been away from home and avoided circumcision have been forcibly circumcised at advanced ages on return home.

Circumcision is practised in many parts of Africa, but unevenly since some peoples have never had it. Since it was done in ancient Egypt, and is still compulsory to Muslims and to Ethiopian Christians, it is not surprising that the practice is widespread in the African tropics. However, differently from the infant circumcision of Judaism and Islam, African circumcision often takes place in adolescence and is part of rituals of initiation. Traditionally it is performed without anaesthetic and it is a test of manhood for a boy to endure pain among his companions. Among the Xhosa it is done when the chief's son is ready to take part in ritual with other youths, and there are distinct groups for eating and drinking among those who have been recently circumcised, others some years ago, and others long ago.

Circumcision often took place in the cold winter season, when the wound would heal better, and groups of naked boys huddled together for warmth. Nowadays it may be done in the summer when boys are on holiday from school, and they return more seriously as adults. Missions, in South Africa at least, often opposed circumcision and even called it 'an unforgivable sin' and expelled such boys from church. One reason for this opposition was a rumour that the boys were taught about sex and how to perform the sex act, though defenders of tradition said that sex and women were taboo during the circumcision period.

At the completion of circumcision, Xhosa youths were brought out of the operation huts, which were then burnt; they were anointed with butter fat and became legal participants in religious ceremonies and rulers of tribal affairs. Some missionaries spoke of the rituals as 'evil', 'barbarous', or 'degrading', but some modern African Christian clergy, themselves circumcised, affirm the morality of the rituals and their effect in making the youths responsible members of the society. Other clergy oppose circumcision as heathen, immoral, and unnecessary. But there is a resurgence of African customs, as there is of Islam and other religions against what seem to be destructive influences from the West. In Africa there are likely to be syntheses of Christian belief and African custom, as already appear in the Independent churches and notably in the Jamaa described below.

Girls had longer times of initiation than boys, following the first menstruation. Even after adolescence menstruation was regarded as unclean or dangerous, and such belief is held widely in Africa, as in some other countries, and even in some churches when women are not supposed to attend church or Holy Communion at such periods. A menstruating woman was called 'in retreat', or as 'having no hands'. She often had to stay in a special hut, and it was taboo for her to eat in company. Although an uncircumcised Xhosa man could not raise a family, he was allowed to herd the family cattle and milk the cows. But women, because of their periods, were regarded as defiled and liable to affect the cattle adversely. 'Such is the extent of male chauvinism in Xhosa society', comments one of their clergy.

Young girls were taken to initiation huts to be instructed in sex by an older woman, with applications of medicines to enlarge the vagina. They were kept in seclusion for several months, receiving instruction, and finally brought out fatted and decorated, to show their marriage value. People who lived near the sea often had customs of washing as part of the purification. In descriptions of initiation instruction among the Ndembu of Zambia, detail has been given of sexual techniques imparted to the girls by older women, with different positions for 'the dance of the bed' and 'the sitting dance', with such instructions being called 'sense' or 'women's wisdom'.[7]

Marking the body with cicatrices is a widespread African custom, found in many tattoos, and at times followed by infection so that the scars are grossly exaggerated. In West Africa sacrifices may be offered at the time of marking the

[7] V. W. Turner, *The Drums of Affliction*, 1968, p. 248.

scars to the guardian spirit Eshu or Legba, or to the god of iron whose knife is used, and dances and festivities follow the operation and completion of instruction. Some marks have little or nothing to do with sex, notably the facial marks which may indicate tribal allegiance, and particular marks of special cults. Patterns of scars are made on the chest, belly, and back, and especially inside the thighs and outer genitals. These are said to enhance the erotic zone, the patterns giving aesthetic pleasure in sex play, and they have been called 'a sort of erotic braille', intended to 'catch a man' by increasing his sexual enjoyment.

Female circumcision, or more properly clitoridectomy, is practised in many parts of Africa, though not in all. It involves cutting off the clitoris, and sometimes even removal of part of the labia minora, with the apparent aim of making sexual penetration easier for the man and removing any opposition or rivalry in intercourse from the woman. The Dogon myth of cutting down the clitoris, given above, suggests that the sexual rivalry of man and woman copies that of heaven and earth.

There is no general physiological justification for cliteridectomy, or the wider operation, such as might be adduced at times for male circumcision. Its purpose seems to aim at ensuring male pleasure and dominance, without considering the women, and men have admitted to thinking that their wives were imperfect or too closed up without it. Those who imagined that Africans traditionally lived in simplicity and enjoyed all sex naturally, can have had no idea that women, half the population, were deprived in many tribes of pleasure in sex. The effect was to subdue women, and certainly to make them suffer unnecessarily, and the anthropologist R. S. Rattray thought that this was one of the few instances where governmental prohibition was needed to protect women. The practice was not universal in Africa, it has probably declined somewhat today, though not enough, and it had little religious significance.

DOWRY AND POLYGAMY

After a long process of initiation the formalities of marriage may be relatively short, and social rather than religious. As in other continents, African marriage was regarded as a contract between two families rather than a romance between individuals, though in modern times personal choice has increased.

Betrothal was often arranged in childhood or early adolescence. Ba-ila parents would go round the villages looking for marriageable girls and saying, 'We are looking for a pot', and in return they would offer male symbols in the form of several hoes. These were not reckoned as part of the dowry but rather as retainers until formal gifts were exchanged.

The terminology used to describe such gifts has been much debated. The English word 'dowry' has meant the portion a woman brought to her husband, whereas in African usage it was the reverse; the husband or his family paid the price. Therefore the term 'bride-price' has been used, but this might give the misleading impression that the woman was bought and sold, and this would be repudiated since there are other words for such transactions and buying a person refers to slavery. A wife was not bought, and her husband did not have proprietary rights over her as he would over a slave. So the term 'bride-wealth' was invented, though 'dowry' could be used with the understanding that it was the husband who dowered.

The amount of bride-wealth differed greatly, from place to place and from rich to poor. Among cattle-rearing people their animals were the most important element in the exchange, and the basic purpose of the payment was to ensure stability of the marriage, since a breach of the contract might involve repayment of the bride-wealth.

Among the Ba-ila marriage came at the conclusion of the initiation, when the girl was taken direct to the bridegroom's house. The man was seized, sometimes reluctantly, and taken home and when it was dark the couple slept together and the union was consummated. The bridegroom took strings of beads equal to his wife's height and hung them on the bedpost, and put a hoe in the fireplace. In the morning an old woman who had prepared the room took these objects as her reward and proof of consummation. Then the family came and placed bread between the couple, which they ate in a communion meal, the woman breaking off a piece and giving it to the man, and the man giving her a piece and also bestowing a new name on her. The family joined in 'the eating of bread', and further gifts were brought. These and further actions formed a true 'rite of passage', in which the couple passed from their childhood state to the adult condition, with sexual union, new name, and exchange of food.

Fifty years later the marriage ceremonies of the Ndembu showed some comparable characteristics. At the end of the initiation the girl went to her betrothed's house, and he had been given aphrodisiac medicines to produce strong erections. The bride was carried over the threshold into the hut backwards, to avoid evil influences and barrenness. Two arrows which the parents had exchanged at betrothal were fixed in the ground at the end of the bridal bed. The couple were supposed to have intercourse as many times as possible, directly and without sexual play, but this was meant to be a test of virility and the rule to be *coitus reservatus*. Presents were left for the instructress, who entered before dawn with washing water to discover whether matters had gone satisfactorily. Though either partner might be lacking, it seems that the onus of proving sexual adequacy lay on the man. After washing the couple received visitors, and that night they could indulge in fore play and gain stimulation from the bodily cicatrices since sexual ability had been proved.

Similar patterns might be adduced from all over the African continent, though many variations have developed in modern times. Marriage is still a contract between families, but divorce seems to be increasingly frequent in only partially-completed marriages. One reason for this is the excessive amounts of bride-wealth that have been demanded and that make for long deferment of formal weddings. This applies particularly to the marriages of Christians in Africa, or of those who wish to be married under a European ordinance. Thirty years ago K. A. Busia contrasted the expenditure in Ghana between Customary and Ordinance Marriages. The former involved gifts of rum on inquiry and agreement, head money and contributions towards puberty rites, and customary presents of clothing for the bride, with small gifts to her father and brothers.

Under Ordinance Marriage there was enormous expenditure on engagement, European and native dress, hardware including a sewing machine, toilet, bridesmaids' dresses, licences, church fees, hire of cars, wedding cake, ring, cost of reception, and photographs. The wife's trousseau to be provided included dresses, hats, shoes, stockings, underwear, chemises, vests, brassieres, roll-on girdle, gloves, handbags, veil, dressing gowns, nightdresses, handkerchiefs, and bouquet. Even the bridesmaids had to be given two sets each of dresses, hats, shoes, stockings, underwear, and gloves. Busia pointed out that such crippling, and sometimes bankrupting, expenditure obscured the concept of marriage. It became much more of a purchase and less of a contract between families. The basic requirements of agreement and exchange of tokens, which made for the stability of the union by providing communal support, were distorted by a gigantic commercial transaction.[8]

Christian missions in Africa, after initial hesitations, generally disliked Customary or Native Marriage, and preferred European-based Ordinance or Statutory Marriage, for which they often pressed. There were 'heathen' customs, not to mention strange sexual practices, involved in Customary Marriage and they seemed informal, arranged by families instead of individuals, and essentially polygamous. To be on the safe side the missions came to insist, or try to, on marriage in a church and conducted by a priest or minister. It is ironic that such insistence by Protestant missionaries followed the Counter-Reformation Council of Trent of the sixteenth century; for previous to that Council Christians had accepted European customary or 'native' forms of marriage, which did not require either a priest or a church.

Christian teaching on church marriage in Africa has often done little to improve the status or stability of marriage in general. Members who contracted a Customary Marriage

---

[8] K. A. Busia, *Social Survey of Sekondi-Takoradi*, 1950, pp. 143 ff.

were likely to be put 'under discipline', a temporary ex-communication. In South Africa, says one modern authority, 'the result has been that nearly all married Christians have been excluded from communion at one time or another in their lives, since a very small percentage marry in church in the first instance, and even the percentage of those who fix up their marriage in church afterwards is relatively small.'[9] The same might be said, in varying degrees, of other places, but there have recently been moves to have Customary Marriage recognized by the churches.

One of the major objections to Customary Marriage was that it presupposed polygamy, or at least did not exclude it. Early missionaries did sometimes recognize the first union, but the husband would be required to put away all his other wives. Further, baptism of converts was generally refused to the man, and even to his wives and children, until he had put away all the wives but the first. Africans were not slow to point out that polygamy was practised by some of the heroes of the faith in the Bible, and some discovered that it had been discussed at the Reformation and apparently allowed by Luther. Missionaries found that their home churches had no formal regulations on the subject, but they did not delay in inventing them and banned even enthusiastic polygamous converts from baptism and office.

Polygyny, male polygamy, has been practised in many countries of the world, but polyandry, female polygamy, is much more rare and does not seem to have occurred in Africa even in matriarchal societies. The strongest argument against polygamy may be that it seems to imply the inferior status of women, their use as property if not as chattels, when chiefs and rich men take extra wives to enhance their prestige. The education of women and economic advancement seem to be reducing the frequency of polygamy, and they also tend to reduce the numbers of children since higher education is expensive.

[9] B. Kisembo and others, eds., *African Christian Marriage*, 1977, pp. 7 ff.

FERTILITY

The tutelary spirit of the city of Ibadan in Nigeria was traditionally held to be a goddess of fertility. She was said to have two huge breasts like large waterpots, so big that sixteen children could suck them at the same time. She was the mother of fertility, and women went to her shrine to pray for children at any time of the year and they made sacrifices at her annual festival. Seventy years ago a European resident of the town said that when the chiefs were asked what emblem they would like to figure on medals that were being given away, they unanimously selected this goddess. A photograph was taken of a fine-looking woman, her breasts exposed and arms raised towards heaven as if to welcome children. It was a picture of fruitfulness.

There was an annual festival of the goddess, held at the time when the land was driest and rain was needed to quench its thirst. It was an occasion for Saturnalia, when no trade was done and anyone coming into the town might be robbed. The town chief threw money into the crowds of people who paraded the streets, and bands of men and women, boys and girls, wandered round the town, throwing open their clothes as if to invite copulation. Wives were said to tell their husbands that they were going to play and picked out the men they fancied.

That was when Ibadan was a country town, but the festival is still held on the modern tarred streets and in the shadow of giant stores and skyscrapers. The day is announced in the press and no fire should be lit or shopping done. The priest in charge of the cult, the 'worshipper of the hill', goes to the shrine which is in enclosed woodland to sacrifice, and on return he visits some of the principal compounds of the city. He wears his hair braided like a woman and has a woman's turban, but otherwise he is in male dress. Women prostrate themselves before him to receive blessings. Groups of people roam the town singing licentious songs and brandishing images of phallic representations and cleft sticks with bunches of pubic hair, or modern pornographic photographs. Small girls carry banners on which are roughly printed the names of male and female sex organs. Many of the shouts and songs are sexually suggestive, especially about people who are unpopular, politicians or foreign tribes. The utterance of taboo words serves as a safety-valve for pent-up feelings, as well as encouraging sexuality.

The festival continues, but with criticism from some educated and religious people. Newspapers have denounced the 'smutty choruses' and 'equally shameless songs directed against the female sex', and have demanded governmental action to 'stop the nonsense'. Christians and Muslims have tried to have the festival prohibited or expurgated, though with little success. But attacks on women, in which some had been stripped, brought swift action by the police and imprisonment for the offenders.[10]

Similar annual festivals of fertility have been held in many other parts of Africa. In neighbouring Porto Novo there is a yearly ceremonial purification, in which all corners of the town are visited. Bands of women open their cloths to passing men and make actions of copulation with inviting words. At the new year on the Ghanaian coastline formal ceremonies and sacrifices culminate in thousands of people dancing for the 'marriage' of the chief god, in which he is hailed as a loving being and giver of fertility. There is open sexual licence, and in the past the sex act was often carried out in public. That is not allowed today but men have the right to embrace any woman ceremonially, and while it may be decorous on the surface there is an underlying licence. Extraordinary sexual meetings are said to be arranged beforehand and held privately, but nobody should object if they are discovered or if claims for adultery are made.[11]

Among the neighbouring Ashanti it was said that redress for adultery or seduction could be claimed once the days of the annual festival were over. Furthermore, much of the apparent licence was formal rather than actual. A man could say to any girl, 'Fire a gun at me', and she was expected to whisk off her clothes in response. But since girls wear strings of beads round their waists and a little red cloth tucked into it at the front and back, the effect might be no more than it

[10] G. Parrinder, *Religion in an African City*, 1953, pp. 12 ff.
[11] M. J. Field, *Religion and Medicine of the Gā People*, 1937, p. 54.

was in a European carnival when a man asked a girl to unmask and kiss him.

Nakedness, partial or total, was not shameful in traditional tropical Africa, and it was more sensible in such a climate than the heavy clothes that were often later adopted. Even a few years ago many girls and women were bare to the waist for much of the day, and only education and modern respectability has clothed most of them. Among some tribes, like the Somba of northern Dahomey, the women went completely naked or with bunches of leaves over their genitals, and the men only wore penis-sheaths made of long thin gourds to emphasize their sex and suggest constant erection.

The need to encourage fertility was very great in tropical Africa, owing to the high rate of infant mortality. It is universally agreed that many babies died in past ages, though virtually no figures are available to indicate the percentages, and the death rate is still high in places away from towns or good hospitals. Children were needed and desired, and well cared for if they survived. Children, especially sons, would perpetuate the family, and ensure that offerings continued to be made to the ancestors. Hence there were both social and religious demands for fertility, and disease and impotence or sterility needed to be fought with all the weapons that were available.

TABOOS

Despite his claim that the Ba-ila indulged in sexual relations on every possible occasion, Edwin Smith mentioned many of the taboos on times and places when coition was banned or dangerous. These ranged over such wide areas of life that sex might be hedged around like the divinity of a king.

Various occupations in Africa have had sexual taboos during important periods. Hunters and fishermen were forbidden to have intercourse with their wives or other women during particular excursions, and iron-smelters were in a state of strict taboo when at their kilns in the bush or even on visits to the village during smelting work. In many parts of Africa it was thought that, if a hunter's wife at home committed adultery during her husband's absence, he would be injured in the chase or killed. Having sexual intercourse on the ground in the bush was regarded as abhorrent to Asase Yaa, the earth goddess of the Akan of Ghana; the couple would become deranged and the deity who was believed to make the land fertile would strike it with barrenness.

Medicines usually had taboos associated with them, things patients had to refrain from doing lest the remedies might not work, and the great power of sexual intercourse could neutralize other forces. A man who kept on pursuing women while under doctor's care could be called 'the village dog'. Even a non-medical charm might carry taboos of intercourse at particular times, if it was to be effective.

There were common taboos on sexual intercourse during menstruation and pregnancy. Pregnant women might not have intercourse with other men than their husbands, and men who tried to seduce them would be liable to heavy damages. Similarly the husband might not have intercourse with other women at this time, except with his legitimate wives. During pregnancy intercourse might be allowed during the early months, but forbidden in the last month or two lest the husband 'pierce the child's fontanelle'.

There are widespread taboos on intercourse between husband and wife after the birth of their child, and this has lasting effects. Most frequently the mother might not have intercourse while she was suckling the child, lest she conceive and her strength be drained away. The Ba-ila said of one who broke this taboo, 'She has jumped over the child, it will waste away.' Since babies are often breast-fed up to two years of age, if not longer, this is a very lengthy period of sexual abstinence both for the wife and the husband. Very often this is the time when the husband takes a second wife, and births from the two women may be spaced out accordingly. The incidence of polygamy may be reduced nowadays by post-natal care and artificial feeding of children.

There are always prohibited degrees of marriage and traditional tables of affinity of sexual relationships. The most abhorred connection is that of incest, both with close members of the family and generally in the wider clan. A taboo on incest always remains taboo, and even at saturnalian carnivals and licentious dances people who are related may not take each other as partners. Since many African peoples practise exogamy, marriage with persons outside the clan, the incest taboo is widespread. Members of different clans in the same village may marry, but no regular or irregular intercourse may be held with members of the same clan even if they live far apart. If nevertheless such a relationship did occur through ignorance, it would be dissolved or allowed to fade away on discovery. The ancestors would be offended, would not heed the prayers of the other partner, and would prevent or damage the birth of children.

In communities where children are needed and loved, there are strong taboos on abortion as bringing uncleanness to the mother, danger to the community, and especially in destroying young life. Masturbation generally is tolerated, in the absence of a theory against it. Homosexuality occurs, and there are occasional men who dress as women and do women's work, but they may be regarded as strange or religiously inspired.

The general attitude to sex in Africa is life-affirming, and it has more in common with the Bible and the Qur'an than with world-renouncing religions. A principal motive for marriage is propagation of the race, in view of the hazards to the health of children and older people. Africans, or black people in general, are sometimes said to be more sensual than other races, and a white writer in South Africa declared, 'Africa, your gods are sex.' This is not true of all African gods, but it comes strangely from a country where the large 'coloured' population is the result of the sexual onslaught of white men upon black women, and the present strictness of laws against miscegenation proves that the temptation is still strong.

CHANGE AND DECAY

If Africans had their taboos, Europeans in Africa had other taboos or complexes. Where nudity was the common practice they often saw degradation or immorality, or tried to avoid being tempted by it, and some of them sought to clothe the tropical maidens in poor imitations of their own female dresses.

Mary Kingsley, the independent explorer of West Africa in the 1890s, wrote:

> I must break out on the subject of Hubbards; I will promise to keep clear of bad language let the effort cost me what it may. A Hubbard is a female garment patronized by the whole set of missions from Sierra Leone to Congo Belge . . . Their formation is this—a yoke round the neck and shoulders fastens at the back with three buttons—two usually lost; from this yoke protrude dwarf sleeves, and round its lower rim, on a level with the armpits, is sewn on a flounce, set in with full gathers, which falls to the heels of the wearer . . . These garments are usually made at working parties in Europe; and what idea the pious ladies in England, Germany, Scotland and France can have of the African figure I cannot think, but evidently part of their opinion is that it is very like a tub. [12]

Mary Kingsley thought that such unnatural clothes helped to produce 'the well-known torpidity of the mission-trained girl', and Dickens had been even more biting in his criticism of this kind of work in *Bleak House* and *Pickwick Papers*.

Since those days much has changed in Africa; great ports, industries, offices, and shops have been built, and many Africans dress much like people in the rest of the Western world. Many African Christian women wear brightly coloured national clothing, but half-nudity is rare. As in much of the rest of the colonized or industrialized world, African customs have been under attack and decline, and there have been great changes in religion and morality.

It would be easy to lament the past, as if everything had been perfect then, and to overlook the sufferings caused by clitoridectomy, infant mortality, polygamy, and customs from which people, and especially women, suffered. Similarly it would be foolish to ignore the achievements of

---

[12] M. H. Kingsley, *Travels in West Africa*, 1897, pp. 220 f.

the new religions in educational and social advance, individual choice in marriage, and real partnership and love of some married couples.

Nevertheless there have been large areas where the clash of old and new has led mainly to destruction rather than creation, and this is particularly true of sex and sex education. There have been a few, but very few, attempts to transform and adapt the old initiation procedures, but they have often been abandoned with no substitute made for them. An African clergyman in South Africa said:

Since sex-teaching was done in the initiation schools when the Africans were still heathens, and since all that and the sex-behaviour that went with it now belongs to the bad old ways we have left behind us, they now associate all sex-talks with heathenism, and they think that since they are now Christians it has become something not to talk about, and certainly not with religion and in church. [13]

Mia Brandel-Syrier, in her detailed and moving account of the lives of African Christian women comments that: 'Since Africans have become converted to Christianity, a connection between religion and sex is for most women not only absent but even positively shocking.' Then missionaries lament that Christianity is only skin deep but, she replies, 'the tragic truth is that it is the missionaries themselves who have brought about the divorce' between sex and religion. In traditional Bantu culture sex was one of the most important expressions of 'life strength', the instrument of human immortality, and it had a ceremonial-ritual significance. But, in all honesty, has missionary Christianity even tried to reach the vital parts of African life? 'Maybe it is the very sexlessness of Christian sacred notions which has made the women and girls dissociate sex from the Christian religion.' Yet Christianity could have influenced sexual behaviour if it had been true to its origins, had understood better the life-affirming views of the Bible, and had investigated African religion and morality sympathetically, instead of condemning them outright as 'heathen'.

[13] M. Brandel-Syrier, *Black Woman in Search of God*, 1962, pp. 143 ff., 217.

Yet perhaps it is not too late, for there is both a resurgence of older African customs, and an interpretation of them by independent churches. One of the most interesting of these churches and movements is Jamaa 'family' of the Congo region, which helped to spread Christianity among mine workers and gave it native roots and impetus, following the reaction against European forms of Christianity after independence. Jamaa was founded by a Belgian priest, Placide Tempels, in his disillusionment with traditional missionary methods and his interpretation of African ideas in his book *Bantu Philosophy*. African thought emphasizes power, he said, and Christianity emphasizes love, and the two should be united.

Jamaa is said to have begun with seven couples, the *baba saba*, 'seven fathers', in Swahili. The aim of initiation into the movement was union with God, the *baba* with Mary and the *mama* with Christ. Then came union of the *baba* and *mama*, physical and spiritual, to emulate the union which existed between Christ and Mary. Thirdly, the *baba* and *mama* were united with the priest, who offered their marriage to God and united each with the other. Married love was regarded as the prototype of Christian love, and the deepest encounters were with members of the opposite sex.

It is perhaps not surprising that there arose a heterodox deviation which taught that the Jamaa ideal of unity in love could only be achieved completely through sexual intercourse, between spouses, non-spouses, and even with the priest. Much Biblical interpretation became sexually oriented, including relationships between Christ and Mary, John and Mary Magdalene. The story of the Annunciation to the Virgin Mary has been explained as the physical union of God as husband with Mary.

When Mother Mary said, How shall that be? You do not have the body of a man, how shall I receive you? God said, I am the creator of the male body, how can I lack it? So the life-force of God began to be with her, his fecundity and his being husband. The Holy Spirit entered her and they finished their work.

This sect has been declared unorthodox and its members

excommunicated in some places. Tempels himself was forbidden to teach at various periods and finally summoned to Rome, but the Second Vatican Council had then taken place, with its wider appreciation of other religions and cultures, and it seems that he was cleared of heresy before his death. The Jamaa continues, openly within the Roman Catholic church, and underground, and it teaches love as its central doctrine. Many other African-controlled churches are seeking to adapt the old and the new faiths and moralities to each other.

Chapter 8

ISLAMIC CUSTOMS

In the three great Semitic or exclusively monotheistic religions, Judaism, Christianity, and Islam, there is no sexual duality in the deity. These faiths moved away from old concepts of Father Heaven and Mother Earth, Shiva and Shakti, or Yang and Yin, which elsewhere had given models for human sexuality. Yet because these religions were, in principle, world-affirming and believed that everything came from God, sexual behaviour should have been of religious concern. 'Be fruitful, and multiply, and replenish the earth', was a divine commandment acceptable to all three.

THE PROPHET

That Muhammad was a sensual man, that this was a bad thing, and unsuitable in a prophet, are opinions that have often been held in the non-Muslim world. His faith and practice are central to an understanding of the relations of religion and sex in Islam, but the facts need to be ascertained from the Qur'an and the most ancient and authoritative biographies and traditions. It is essential also to study the matter in the context of the times.

   Muhammad is usually said to have had fourteen wives, or nine wives ,in the strict sense and five concubines. His polygamy was undoubted, by friend or foe, but it was the custom of the time and country. It was and is common for great men in many eastern lands to have several wives, and political alliances have been strengthened by marriages. Similar polygamy was not definitely forbidden to Jews until

151

the Middle Ages, and many Christian rulers have been less than strictly monogamic, for example Henry VIII.

Muhammad does not seem to have been excessively sensual, according to the standards of his time, and an eminent Christian scholar remarks that, 'It is not too much to say that *all* Muhammad's marriages had a political aspect'.[1] Thomas Carlyle a century ago, in 'The Hero as Prophet', noted from Islamic Traditions that Muhammad was ascetical in some ways, living simply and often retiring to the desert for meditation and prayer. 'His household was of the frugalest; his common diet barley-bread and water: sometimes for months there was not a fire once lighted on his hearth. They record with just pride that he would mend his own shoes, patch his own cloak.'

Some Muslim writers, however, in endeavours to endow their Prophet with all the attributes of a popular hero, credited him with extreme sexual energy. The traditionalist Bukhari declared that the Prophet could satisfy all his wives in one hour and had the sexual strength of forty men. Many women, he averred, offered themselves to him, though this did not imply open promiscuity but the offer of marriage without dowry. It may be imagined what medieval European critics made of such tales, for, as a modern writer remarks, 'the facts were so often invented, or else falsified, or just exaggerated.'[2]

The facts are that this Prophet did not marry until he was twenty-five years of age, and then wedded Khadija who was forty years old and had been married twice before. They remained together in fidelity and probably love for some twenty-four years, and she bore all his children but one. Only after Khadija's death, when he was fifty years old, did Muhammad take his second wife, a widow of a leading Muslim who needed protection after her husband had been killed in battle. In the last years of his life, when he had become a prominent figure with a large following, Muhammad took other wives. He died at the age of sixty-two.

Khadija was a merchant woman who employed men to

---

[1] W. M. Watt, *Muhammad at Medina*, 1956, pp. 330 ff and 393 ff.
[2] N. Daniel, *Islam and the West*, 1960, p. 102.

trade for her on a profit-sharing basis. She was so impressed with Muhammad's stewardship and integrity that she proposed marriage to him, which he accepted. She became the first 'Mother of the Faithful' and encouraged Muhammad in his prophetic vocation. When her husband had received his first vision of the angel Gabriel and was struck with wonder in a mountain cave, Khadija sent her messengers after him. He returned to her and 'sat by her thigh and drew close to her'. One reporter said that when Muhammad declared he could still see Gabriel, Khadija made her husband 'come inside her shift', whereupon Gabriel departed and proved by his modesty that he was an angel and not a demon. If Khadija comforted Muhammad with sexual intercourse, that would have provided release from the tension caused by his supernatural visions. Khadija was the first to believe in Muhammad's message. She had a cousin Waraqa, an old man who was a Christian and had studied the scriptures. When Khadija told him of her husband's experiences, he said, 'If this is true, he is the prophet of this people.'

Only one of Muhammad's wives had not been married before. That was Ayesha, his third wife and daughter of his chief follower and the first caliph, Abu Bakr. Ayesha was nine years old and she continued to play with her toys, and may have remained with her mother for some time. Later, on an expedition, Ayesha was somehow left behind and returned home in the care of a young man. This was before the time of veiling for the Prophet's wives and tongues wagged, but there seems to have been no clear evidence of misconduct and the incident came to be referred to as 'the affair of the lie'. Ayesha developed quickly into a fine and witty young woman and probably became the dominating wife. After Muhammad's death she came into conflict with his cousin and son-in-law Ali, when her army met his at 'the Battle of the Camel' near Basra, named after the camel on which Ayesha rode. She was captured but sent back in dignity to Medina.

Muhammad next married the daughter of his second chief follower, Umar the second caliph-to-be, and Umar also married Muhammad's granddaughter. The third caliph-to-be,

Uthman, married two of Muhammad's daughters, and the fourth caliph-to-be, Ali, married his daughter Fatima. Thus these marriages served to unite the Muslim community, though they did not provide an undisputed succession.

Muhammad's marriage with Zainab, wife of his adopted son Zaid, however, has occasioned critical comments. Zainab was Muhammad's cousin and had apparently been married to Zaid against her will. One day Muhammad called at Zaid's house and he was out and Zainab was in a loose and scanty dress. The Prophet refused to enter but went away exclaiming, 'Praise be to God, praise be to the Manager of Hearts.' Zainab told her husband about this cryptic utterance and Zaid went at once to Muhammad and offered to divorce his wife, but was told to keep her. Zainab was an ambitious woman and she made life difficult for Zaid, he divorced her, and after the due waiting-period Muhammad married her. This action was justified by a special revelation in the Qur'an: 'When Zaid had had all that he wanted of her, We [God] married her to you, in order that there should not be any feeling of blame upon the believers, in regard to the wives of their adopted sons' (33, 37). Non-Muslim critics have supposed that this affair proved the sensuality of Muhammad, but his contemporaries were only concerned that the marriage was within forbidden degrees by marrying the wife of an adopted son. The Qur'an showed that there was to be a break with this custom of the past.

Four years before his death Muhammad married Mariyah (Mary), a Coptic concubine presented to him by the Christian ruler of Egypt. She bore him a son Ibrahim, to the great joy of the parents, but the child soon died. There was some jealousy among the other wives over Mariyah, as a foreign and no doubt sophisticated woman, but such strains were common in polygamous households. If Muhammad's aim in taking these numerous wives was to have children to carry on his family, and perhaps to provide a prophetic dynasty, he was even less successful than Henry VIII. His sons all died in infancy, and only one daughter survived him by a year, Fatima wife of Ali.

By the time of his early death, Muhammad had become an important religious and political figure, known in most of

Arabia and some regions beyond. He was the founder of a new religion, parallel to Judaism and Christianity, and recipient of a scripture, the Qur'an, which compared with the Jewish Torah and the Christian Gospel. Muhammad was undoubtedly a man of deep religious feeling, firm principle, and sexual experience. His family life has been taken as a model by successive generations, and modern apologists have claimed that as a married man and a father Muhammad was a more suitable pattern for ordinary believers than celibate or world-renouncing religious leaders.

MARRIAGE IN THE QUR'AN

A verse of the Qur'an distinguished between Muhammad's wives:

> O Prophet, We have made allowable for you those wives whose dowries you have paid; those whom you have taken of the booty which God gave you as property; the daughters of your paternal and maternal uncles and aunts who emigrated with you; and any believing woman who offers herself to the Prophet if he wants to marry her; this is special to you and does not apply to the believers (33, 49).

The first group consisted of wives in the strict sense for whom dowry had been given. The second group would be slave-concubines taken in war or presented as gifts. The third grouping permitted marriage of cousins, perhaps with the limitation to those who emigrated with Muhammad to Medina, and Zainab would presumably be included here. The last group was of Muslim women who offered themselves without dowry. Probably various women and their families were anxious to claim a matrimonial relationship with the Prophet, and some may have contracted a temporary union with him, though none seems to have had apartments in his residence at Medina as his regular wives did.

There were numerous verses in the Qur'an which dealt with the relations of the sexes. One of the most often quoted (4, 3) seemed to limit, or perhaps encourage, men to have two, three, or four wives. This passage was traditionally said to have been given after a number of men had been killed in the battle of Uhud, leaving women and children desolate. The chapter opened with reference to the creation of Adam and Eve, 'Your Lord created you from one individual, and from him created his spouse, and from the two spread abroad many men and women.' Then the question of the orphans was raised, and the care of their property which unsympathetic guardians might take and keep in their control by refusing to marry the orphaned girls. So, 'Give orphans their property; do not substitute the bad for the good, and do not consume their property in addition to your own—this has truly become a great crime.'

The injunction to plural marriage followed: 'If you fear that you may not act towards orphans with equity, then marry such of the women as seem good to you; two, or three or four . . . and give the women their dowries as a gift.' But if such polygamy would cause trouble, 'if you fear that you may not be fair [to several wives], then take one only.'

This verse did not perhaps mean that the guardians should marry their wards, but rather that all girls should be properly married, with dowry, as soon as they reached the right age. It has often been taken to restrict the hitherto unlimited Arab polygamy to four wives at most, though there would be no limit to the number of concubines. Polygamy, we have seen, was common in India, China, and Africa, but in Islam it had this limited scriptural justification. However, in modern times some Muslim apologists have claimed that, while polygamy was permitted in olden days, and under special circumstances, yet it is impossible to act fairly or 'with equity' towards several women, and men should therefore marry one only. Thus the Qur'an is quoted in favour of monogamy.

Marriage in the Qur'an, as among many peoples, was for the procreation of children and was recommended for everyone in the right conditions: 'Settle the unmarried among you in marriage, also the upright among your male and female slaves; if they are poor, God will enrich them of his bounty . . . And let those who do not find the means to marry restrain themselves until God enriches them of his bounty.' (24.32)

Marriage was lawful for a Muslim with a Jewish or Christian woman, but it was not right for a Muslim woman to marry a man of another faith. Marriage with pagans was strictly prohibited until they were converted: 'Do not marry idolatresses until they believe; a believing handmaid is better than an idolatress, even though you admire her.' Similarly women were told, 'Do not marry idolaters until they believe; a believing slave is better than an idolater, even though you admire him.' (2,220)

Sexual intercourse was encouraged but it should be preceded by an act of piety: 'Send forward something for yourselves, and act piously towards God and know that you are going to meet him.' (2, 223) This has been interpreted as commanding the utterance of a pious phrase before sexual intercourse and giving a warning that we shall meet God at the judgement. That done, however, women were regarded as furrows or cultivated land. The metaphor of the furrow would be a development of the primitive notion of coition as the sowing of seed. 'Your women are like furrows to you, so come to your tillage as you wish.' The phrase 'as you wish' has been taken to mean, either have intercourse at any legitimate time, or in reference to various postures in copulation. It was said that 'unnatural practices' were excluded by the words, 'Come to them as God has commanded you', in the previous verse.

Intercourse during menstruation was strictly forbidden: 'They will ask you about menstruation; say, "It is harmful, so withdraw from women in menstruation, and do not come near them till they are clean . . . God loves those who keep themselves pure."' (2, 222)

There were strictly prohibited degrees of marriage, to blood relations on the mother's and the father's side. In pre-Islamic days men could marry their father's wife and also two sisters together, but both were forbidden to Muslims:

Do not marry women whom your fathers have had in marriage; except so far as it is already past . . . Forbidden to you are your mothers, daughters, sisters, paternal aunts, maternal aunts, brothers' daughters, sisters' daughters, those who are your mothers or sisters by suckling, your wives' mothers, wives' daughters, sons' wives, or two sisters at the same time, except so far as it is already past.' (4, 26f)

Marriage and the family were reorganized by Muhammad to give priority to paternity and patrilineal descent, and earlier Arab bars to marriage within blood-relationship on the mother's side were extended to the father's as well. Some of the Arabs before Islam had followed a system of kinship which regulated marriage and descent through the mother. There were also forms of polyandry, in which a woman had several husbands and the physical fatherhood was neglected. The permission for a Muslim to have four wives may have been intended to limit a woman to one husband at a time, as well as to ensure the marriage of surplus women after male deaths in battle. Under this Islamic system the physical paternity of the child would always be known. Polyandric women were to become monandric under Islam with the consent of their families: 'God knows your belief, you are one people, so marry them with the permission of their households, and give them their dowry, but reputably as being now under ward, not committing debauchery or receiving paramours.' (4, 29)

There was an Arab custom called *mut'a*, 'temporary marriage' or 'marriage of pleasure' which was contracted for a fixed period on rewarding the woman. This seems to be referred to in the Qur'an: 'And further, you are permitted to seek out wives with your wealth, in decorous conduct but not in fornication, but give them their reward for what you have enjoyed of them in keeping with your promise.' (4, 28) Such temporary unions may have been tolerated as old practices, though the orthodox explanation of the verse above refers it to normal marriage. Temporary unions were prohibited by the caliph Umar, and the practice was only later permitted among a small section of the Shi'a.

Islamic divorce was carefully regulated, for men. There had to be an interval of three or four months between the declaration of divorce, accompanied by separation, and the actual dissolution of the knot of marriage. 'Divorced women wait by themselves three courses, nor is it permissible for them to conceal what God had created in their wombs, if they have believed in God and the Last Day.' (2, 228) If they were pregnant, they should not hide it but stay with their husbands.

Within the three-month period the way should be kept open for resumption of sexual intercourse, if both parties wished for it: 'Their husbands have the best right to restore them within that period, if they wish to set things right'; but for women also, 'in reputable dealing they have the same right as is exercised over them.' (2, 228)

Divorced women should be sent away with kindness, and it was not permissible to take away their dowries. A divorced wife might not be married to the same husband more than three times, unless marriage with another man had intervened and had been followed by divorce from him: 'Then if he divorces her, it will be no fault upon them that they return to each other, if they think they can maintain the bounds set by God.' Although it was largely the man's prerogative, there was an insistence on the duties of husbands to wives to 'retain them reputably or send them away reputably'; not to 'retain them by compulsion' and not to 'vex them'. These instructions were rounded off with the words, 'that is an admonition to whomsoever of you believes in God and the Last Day; that is more innocent and purer for you; God knows though you do not know.'

Indecent conduct was to be punished in several ways. 'Those of your women who commit indecency, call four witnesses against them, and if they testify confine them to their houses until death complete their term of life.' (4, 19) The following verse mentions 'couples who commit it', and this has been taken as condemning homosexuality, though it could just as easily mean immorality between couples of opposite sex. Elsewhere, more severe punishments were decreed for fornication, both the man and the woman, 'scourge each of them with a hundred stripes, let no pity affect you in regard to them.' (24, 2) Immoral people attract each other, 'Bad women to bad men, and bad men to bad women.'

In earlier centuries, though less today, non-Muslims tended to regard the picture of the early Muslim community that appears from the Qur'an and the Traditions as one of 'sanctified sensuality' or 'puritanical licence'. In fact, it seems to have been a society that was emerging from a background in which religion and sexual behaviour were not always closely related, into one in which the prophetic faith was applied to all the elements of life. In the growth of a distinct religious community adjustments had to be made for new situations, such as the state of widows or orphans after warfare, but the aim was to regulate behaviour by the will of God.

SEX IN THE TRADITIONS

The Traditions have been revered next to the Qur'an itself throughout the Muslim world. They are Hadith or Sunna; the Hadith being a story or account of what happened, while Sunna meant the way or custom which was practised. The Traditions were basically about matters relating to Muhammad and the customs followed by him and his immediate followers. How reliable such narratives and teachings were it is hard to say, but while believers wished to model their lives on that of the Prophet, the great expansion of Islam in the early centuries meant that adaptations had to be made to new conditions and Traditions were provided to justify them. In the even greater changes of modern times many Traditions have been criticized and some would abandon them, yet they furnished patterns of behaviour for centuries.

Circumcision was not mentioned in the Qur'an, but it was later said to be founded upon the Customs of the Prophet and it became Sunna, traditional for all Muslims. According to some authorities it was equally obligatory for males and females. Circumcision was practised in pre-Islamic Arabia and by the Jewish neighbours of early Muslims; Muhammad is said to have remarked that Abraham was circumcised when he was eighty years old, and a later legend maintained that Muhammad himself was born circumcised. A Muslim child may be circumcised seven days after birth, or between three and seven years of age in Arabia and Egypt, and elsewhere up to the thirteenth year. Boys are circumcised with great pomp, but girls without festivities. Clitoridectomy was Sunna as far as it involved removal of the tip of the clitoris, and further practices were excision of the clitoris and labia minora, and infibulation which included excision and then sewing up the vagina. Modern western sexologists hold that all orgasms in women are caused by clitoral stimulation, and therefore clitoridectomy not only causes pain but deprivation, hence some modern Muslim doctors and educated women in eastern Mediterranean countries have urged the abolition of clitoridectomy in all its forms, but there is often great social pressure on parents to maintain the operation on their daughters. [3]

Purification or washing after many occupations was important to Islam, and Muhammad was said to have remarked that 'being purified is half of faith'. Many sayings were reported about the methods of performing excretory functions and ablutions afterwards, and the detail recalls the minute regulations of the Pharisees. The caliph Umar was reported as saying that a man must perform ablution after kissing or touching his wife, because these actions were connected with sexual intercourse. But Ayesha, who had a sharp tongue, said that Muhammad used to kiss one of his wives and then pray without performing ablution; but this was disputed. [4]

Washing was said to be necessary if a man lay on his wife but did not penetrate her, though another Tradition said that washing was only needed when there was an emission. Ayesha said that the Prophet washed after a seminal emission and then performed ritual ablution as for prayer, but she was also reported as saying that she and the Prophet would wash from the same vessel which stood between them and he would get ahead, so that she exclaimed, 'Give me a chance.' [5]

It was reported that the Prophet saw a man washing naked in a public place, so he mounted the pulpit and after praising God he said, 'God is characterized by concealment, so when

---

[3] See S. Hite, *The Hite Report*, 1977, pp. 179, 271 f.; W. Masters and V. Johnson, *Human Sexual Response*, 1966, p. 59 f.

[4] *Mishkat al-Masabih*, tr. J. Robson, 1960, 2, 3; 3.6   [5] Ibid., 3.6.

any of you intends to wash, he should hide behind something.' Another report said that men should avoid being naked because the recording angels were always present, except when men were excreting or having intercourse. Muslim practices of washing and excreting in private and without stripping naked, were based upon this and similar Traditions. Urination was forbidden in mosques and public places, and after treading on something unclean earth was used to purify the sandal. [6]

A woman is said to have asked Muhammad how to wash after menstruation and he told her to take a piece of cotton with musk. When she demanded details of the procedure he exclaimed, 'Praise be to God! Purify yourself with it.' Then Ayesha instructed her. But the Traditions seem to have been less strict on this point than Jewish or pagan practice, for being told that the Jews did not eat or live with a menstruating woman, the Prophet is said to have remarked, 'Do everything except sexual intercourse.' Ayesha was reported as saying that when she was menstruating her husband would tell her to wrap herself up, but then he would embrace her, drink from the same vessel as her, and put his mouth where hers had been. He would recline on her lap at such times and recite the Qur'an. He would take his prayer mat from her hand and say, 'Your menstruation is not in your hand.' Intercourse during menstruation was regarded as unbelief, but the penalty was only a fine. A constant flow of blood was not menstruation and so prayer could be continued in that condition, while during normal menstruation prayer should be suspended because Islamic prayer involved not only speaking but washing, bowing, and prostrating. [7]

There were many Traditions about marriage, and the Prophet was said to have declared that 'the whole world is to be enjoyed, but the best thing in the world is a good woman.' He disliked celibacy and a Tradition said that 'there is no monkery in Islam.' The Qur'an had spoke of the 'kindness and mercy' of the followers of Jesus, but said of 'monasticism, they invented it.' (57, 27) So the Traditions told Muslim young men to marry to avoid immorality and

[6] Mishkat 13.2.   [7] Ibid., 3.13.

looking at strange women, but if they could not afford it they should devote themselves to fasting, 'for it is a means of suppressing sexual desire.'

The Prophet was reported as saying, 'Marry women who are loving and very prolific.' To a man who had just married a woman who had been previously married, he was reported as saying that he should have married a virgin, 'with whom you could sport and who could sport with you', and to another he is said to have declared, 'Marry virgins, for they have the sweetest mouths, the most prolific wombs, and are most satisfied with little.'[8]

In pre-Islamic and early Islamic days some Arabs practised *coitus interruptus* and it was not prohibited. The Greeks and Persians were said to have practised withdrawal without harm, and their women were also noted to have suckled children during pregnancy without injury. A man was reported to have gone to the Prophet saying that he had intercourse with his slave-girl and did not want her to conceive, and he was allowed to practise withdrawal because whatever God had decreed would come to her. Other Muslims had taken some Arab women captive and wished to practise *coitus interruptus* but asked the Prophet first who was said to reply, 'It does not matter if you do not do it, for every soul that is to be born up to the day of resurrection will be born.' For him faith in predestination was dominant, and this principle might be applied to the practice of birth control today, which many Muslims oppose.[9]

Several Traditions forbade intercourse by the buttocks, that is through the anus, and the Quranic verse was quoted to this effect, 'Your wives are a furrow, so come to your furrow as you wish, but act piously towards God.' The Traditions added that, 'He who has intercourse with his wife through her anus is accursed', and again, 'God will not look at a man who has intercourse with a man or a woman through the anus.' One verse even said that he who had intercourse on the back of his wife, but through the vagina, would have a child with a squint.[10]

Women were to be treated with consideration, the reason

---

[8] Mishkat 13.1 f.   [9] Ibid., 13.6.   [10] Ibid.

being given that they were created from Adam's rib and if you try to straighten a crooked rib you will break it. A wife should not be whipped like a slave and then subjected to intercourse, but when a man called his wife to satisfy his desire she should go to him even if she was occupied at the oven. On the other hand, a wife should not tell others of her marital affairs. Being asked what rights a woman could demand of her husband, the Prophet was reported as saying, 'That you should give her food when you eat, clothe her when you clothe yourself, not strike her on the face, and not revile her or separate from her.'

The penalties for fornication were severe, and it has been seen that the Qur'an decreed a hundred stripes to each offender. But the harshness of the Jewish law seems to have caused problems. Some Jews were said to have come to the Prophet and told him of a man and woman of their number who had committed fornication, and he asked them what the Torah taught. They said that they should be beaten, but another declared that they lied and the punishment was death by stoning. Muhammad asked for the Torah to be brought, and when the verse on stoning was read he ordered that it should be carried out. Another verse, however, said that Muhammad distinguished between the fornication of the unmarried, which should be punished by a hundred lashes and a year's banishment, and that of married people who should be stoned to death. [11]

Several Traditions told of a man, or men, who came to the Prophet in the mosque and confessed that he had committed fornication. Muhammad turned away, but the man came round facing him and repeated the confession. The Prophet turned away again, until the fourth confession, when he spoke to the man and said, 'Are you mad?' When the man answered that he was not, the Prophet asked if he was married, and when he admitted this the order was given to stone him to death. When the stones hurt him the man ran away, but the stoners caught and killed him. Then the Prophet spoke well of him and prayed over him. [12]

It seems that Muhammad was fulfilling the law passed

[11] Mishkat 16.1.　　　　　　　　　　　　　　[12] Ibid.

164

down by Jewish custom, but he tried to temper it with mercy. Ayesha reported him as saying, 'Avert the infliction of prescribed penalties on Muslims as much as you can, and if there is a way out then let a man go, for it is better for a leader to make a mistake in forgiving than to make a mistake in punishing.' Such an attitude anticipated modern objections to capital punishment, on the grounds of possible mistakes and its irreversible nature.

In modern times some Islamic states that attempt to follow strict orthodoxy continue to apply the Traditions, for example stoning for adultery and cutting off hands for stealing. Yet such regulations are not in the Qur'an, as western newspapers often say, but in Traditions which are many and varied and of differing authenticity. How important the Traditions are, and whether they should be followed today, are questions much disputed in Islamic countries. There have been modern calls for 'Back to the Qur'an', rather than to Traditions, and an Egyptian professor claimed that the small value of the Traditions is like a handful of gold in a heap of chaff. Yet in the past the Traditions, and legal interpretations of them by teachers of the community, have governed much of Islamic life, and on sexual matters they have had great influence.

SEX IN LITERATURE

Apart from the Traditions and commentaries by legal exponents of them, there arose a vast Islamic literature on all aspects of life. This developed when Islam spread throughout the Middle East, and its extent can be seen in histories of Arab, Persian, and other literature, which list thousands of works without clear reference to sex which nevertheless appeared in many writings.

After the first four caliphs the rule of the Islamic empire shifted f.. .n Medina to the ancient city of Damascus for the century of the Umayyad caliphate, and then to Mesopotamia for some five hundred years of the Abbasid caliphate. The new city of Baghdad grew from nothing to a world centre of wealth and culture, rivalled only by the Orthodox Christian city of Constantinople with which it had frequent contacts.

The most brilliant period of Baghdad was the caliphate of the famous Harun al-Rashid (786–809). The royal palace with annexes for harems, eunuchs, and special functionaries occupied a third of the city, with tens of thousands of curtains and carpets, luxuries virtually unknown at that time in western Europe. The caliph studied Persian and Indian philosophy, while his western contemporary Charlemagne was trying to learn to write his own name. Many women, at least at the top, lived glamorous lives and the caliph's cousin-wife Zubaida would have at her table no vessels not made of silver or gold or studded with gems. She set the fashion by being the first to ornament her shoes with precious stones. Her rival Ulayya covered a blemish on her forehead with a fillet of jewels, which also became a model for fashion.

The court of Harun al-Rashid provided humorous anecdotes and love romances for the famous Thousand and One Nights (the so-called Arabian Nights), though it was not finished for several centuries. Behind it were many sources, and one was an old Persian book which told of a king who would marry a woman for one night and kill her in the morning. Many stories were added to this, some belonging to Baghdad in which Harun al-Rashid figured, and others based in Cairo, with many Indian, Greek, and other oriental stories of every description.

The Thousand and One Nights opened with an invocation to the Name of God, the Merciful, the Compassionate, with praise to him and peace upon the Prince of Messengers, Muhammad, and prayer and peace upon his people. Then sexual adventures were freely related between all manner of people, high and low. The very first story told of the king's wife being found stretched on her bed and embraced by a black slave, and in a neighbouring kingdom the queen undressing with her men and women slaves and being embraced by a gigantic negro who turned her on her back and enjoyed her. These orgies came to an end, with all being beheaded, and the king ordered his vizir to bring him a young virgin every night, whom he ravished and killed at dawn. This continued for three years till the people fled with their daughters, but the vizir's daughter Shahrazad offered herself, for she was very wise, and said, 'By God, father, you must marry me to this king; for either I shall live or, dying, I shall be a ransom for the daughters of the Muslims.' She began to tell the king stories and left each tale at a point which induced him to spare her life and continue the following night. The theme is women's wiles, 'for whenever one of us women desires a thing, nothing can prevent her from it.'

Although the Thousand and One Nights tells of many sexual adventures, and often shows women enjoying sexual intercourse as much as men, it is by no means a manual of sex, but a reflection of life and of fantasy. There were more specifically sexual manuals, especially in Egypt, and even the theologian and historian Jalal al-Din al-Siyuti was credited, perhaps wrongly, with writing a *Book of Exposition of the Science of Coition*. This opened with the exclamation, 'Praise to the Lord who adorned the virginal bosoms with breasts, and made the thighs of women anvils for the spear-handles of men.' Other erotic works had titles such as the *Book of the Bridal and the Brides*, the *Maiden's Book*, the *Mysteries of Married Fruition*, the *Compendium of Pleasure*, and the *Initiation into the Modes of Coition*. Such works showed how men and women could gain pleasure from sex and, like the Indian *Ananga Ranga*, they provided instruction in variety in sexual methods for married couples, though often they depicted wider adventures with different partners. Indian works like the Kama Sutra were translated and illustrated for use by Muslims.

Islamic works on sex normally opened, like books on almost any other subject, with the praise of God and thanks to him for granting such pleasures to mankind. But it is debatable how far such works went beyond the physical and mental enjoyment of intercourse, to the mystical union which characterized many Indian and Chinese books. An expert on Islamic law states categorically that, 'although a religious duty, marriage is emphatically not a sacrament. There are no sacraments in Islam.'[13] This might be disputed, on a broad interpretation of sacraments, but it would perhaps be going too far in the other direction to include Islamic sexology along with Indian and Chinese in

[13] S. Vesey-Fitzgerald, *Muhammadan Law*, 1931, p. 37.

sacramental or mystical classifications. Yet another western writer says that:

> the instructional manuals of Oriental erotology present love as, in a sense, a sacrament, and look upon the sexual act not only as a means of procreation but as a healthy (even a healing) pleasure. On its highest and purest level they regard coition virtually, and sometimes literally, as an act of worship. An Arab poet has ideally expressed the fundamental reverence underlying their attitude: 'Love enters in through the eyes, which are the doors of the spirit, and then diffuses himself throughout the whole of the soul.' Thus, for the Oriental, orgasm symbolized the ecstasy of the soul possessed by, or in union with God.[14]

But 'the Oriental' should not be so generalized, and there are vast differences between the pantheism of India and the strict monotheism of orthodox Islam. The notion of union with God was abhorrent to many Muslims, who sought only to be surrendered to the omnipotent will. That Islamic mystics did somehow manage to bridge this gulf between God and man is one of the curiosities of the history of mysticism. Reference will be made later to some mystical writings, but Islamic sex manuals generally are to be understood against a monotheistic background rather than a pantheistic one.

In the sixteenth century Shaikh al-Nafzawi of Tunis wrote *The Perfumed Garden for the Soul's Delectation* which was both fantastical romance and sexual instruction. It opened with the general invocation, 'Praise be to God, who has placed man's greatest pleasure in the natural parts of woman, and has destined the natural parts of man to afford the greatest enjoyment to woman.' But it went on to deny that a woman could have any pleasurable feeling until penetrated by the male, and affirming that the man knows no rest or quietness until united with the woman. Coition was described as like two actors wrestling. The lips are the gift of God, and so are the breasts, eyes, belly, and especially

---

[14] A. H. Walton, in Introduction to *The Perfumed Garden*, trs., R. Burton, 1963 edn., p. 40.

'the arena of the combat', which was described in great detail. 'So let us praise and exalt him who has created woman and her beauties, with her appetizing flesh . . . I, the servant of God, am thankful to him that no one can help falling in love with beautiful women.'

The Perfumed Garden continued with statements and stories, the latter resembling tales from the Thousand and One Nights. After brief discussion of physical qualities which women look for in men, there came a vivid tale of the copulations of a court fool, Bahloul, and a vizir's wife, Hamdonna, which she enjoyed as much as he. Then followed a description of desirable women, and a story of a negro who sought to seduce the vizir's wife but was finally mutilated and executed. Further chapters discussed different kinds of men and women, provided a prayer for a pregnant woman to be delivered of a boy, listed eleven positions for intercourse and twenty-five others which were said to be used by the Indians, who have advanced 'farther than us in the knowledge and investigation of coitus', and more movements to be practised during intercourse. Other chapters considered matters injurious to intercourse, thirty-five names for the penis, thirty-eight for the vagina, and names of male organs of animals. Stories illustrated the deceits and treacheries of women, the causes of enjoyment, sterility, abortion, impotence, small penes, bad smells, knowledge of sex of a foetus, and the use of eggs to help coitus. Arabic versions of *The Perfumed Garden* contained a section on homosexuality, which is omitted in most European translations. The whole was rounded off, somewhat like Chaucer's *Canterbury Tales*, with a prayer to God for forgiveness, 'O God, award no punishment for this on judgement day', which confirms that it was rather erotic fact or fantasy than mystical contemplation.

MYSTICAL SYMBOLISM

Pictures of Paradise or the Garden of Eden were given in the Qur'an in physical imagery much like those of Zoroastrian, Jewish, and Christian eschatology. But the Qur'an went further, and on some points, to the satisfaction of believers and the scorn of critics. The blessed in Paradise would recline on couches set with jewels, while round them circled boys of perpetual youth, with goblets and jugs and a cup of flowing wine (forbidden to Muslims on earth), from which they would drink and suffer neither headache nor intoxication.

Most notable were the 'houris' (*hur*), meaning 'white ones' or 'bright-eyed', the maidens the black iris of whose eyes was in strong contrast to the clear white around it. The houris are mentioned in the Qur'an four times and one or two other passages also describe them as companions of the blessed. They are like treasured pearls, 'a recompense for what they have been doing'; they are 'spotless virgins', 'amorous and of equal age', like hidden pearls or ruby or coral, with swelling breasts, deflowered by neither men nor jinns, enclosed in pavilions, and with shy, retiring glances. (55, 56; 56, 22)

These verses all occur in the chapters of the Qur'an which are usually dated in the early period of Islamic life at Mecca. Some non-Muslim scholars have suggested that the idea of the maidens of Paradise came from Zoroastrian or Christian pictures of angelic figures, which often look feminine. But the Islamic houris continue into the world to come the sensual pleasures of this world and the service of virgins to men. Later Islamic literature gave more details of the houris; they were so transparent that the marrow of their bones was visible through seventy silken garments, and they had two names written on their breasts, one being the name of God and the other that of their husbands. But scholarly Islamic commentators said that, although heavenly women had the name in common with their earthly equivalents, yet this was only 'by way of metaphorical indication and comparison, without actual identity'.

In some later Medinan chapters of the Qur'an, with developing theology, there was mention of 'purified spouses' in Paradise, and it is not clear whether these were the houris or the earthly believing wives of Muslims. However, in contradiction to the notion that women do not go to heaven in Islamic belief, there are statements in the Qur'an that believing men, women, and children will enter heaven as families: 'They shall enter Gardens of Eden, with their spouses and posterity who have been upright', and again, 'O our Lord, cause them also to enter Gardens of Eden which thou has promised them, and likewise those of their ancestors and spouses and offspring who have lived uprightly.' (13, 23; 40, 8) Men and their wives will enter the Garden in gala attire and recline on couches in the shade, eating fruit and whatever they call for. 'The self-surrendering men and the self-surrendering women, the believing men and the believing women', the obedient, truthful, enduring, submissive, almsgiving, fasting, God-remembering men and women, 'for them God has prepared forgiveness and a mighty reward.' (33, 35)

Mysticism, which may be defined as union with God or the ultimate reality, did develop in Islam despite the strong contrast between God and man in orthodox theology. It may be, as one writer suggests, that since God alone can annihilate the infinite distance between himself and man, God does precisely that by his omnipotent will. Others have seen Indian influence in the virtual pantheism or identification with God which some Muslim mystics expressed.

The love of God for man, and man for God, which hardly appeared in the Qur'an, soon became a major theme of Islamic mystics. One of the earliest was a woman, Rabi'a of Basra, who remained celibate and was called 'a second spotless Mary'. Rabi'a spoke of God as the Companion and Beloved of her heart, yet while such sentiments might be interpreted as the sublimation of sexual urges, there was little overt sexual symbolism in her teaching.

Much the same could be said of other Islamic mystics who used the language of love, but did not write in the erotic manner of the more frankly sexual romances. Sa'di of Shiraz in the thirteenth century wrote mystical poems on the Garden and the Rose Garden, which might be thought to use symbolism like the Perfumed Garden, but they were exhortations to personal and social piety, with some verses on the infatuation of the love of God and simple illustrative stories. A dancer had her skirt burnt by a candle, and one of her lovers told her not to be distressed because the fire of the love of God had consumed his life.

Perhaps the most celebrated Persian composer of love poems with a spiritual meaning was Hafiz in the fourteenth century, also in Shiraz. He spoke of his beloved, his lady, of love's true fire and its passionate red wine, but whether it was human or divine love there was not the sexuality of the truly erotic books.

For the West it was Omar Khayyam, a Persian mathematician and poet of the twelfth century, who seemed to sing only of love and wine, though in his own country he was regarded as a mystic of the second rank. Looking at pottery he was reminded of love, but also of the passing show of life and the crumbling of it all to clay and dust:

> This Pot was once a Lover in the Chase,
> Like Me, Admirer of a beauteous Face;
>   This Handle fastened to its Neck was once
> A Hand stretched out the Loved One's Neck
>   to embrace.

The most famous verse runs:

> Here with a Loaf of Bread beneath the Bough,
> A Flask of Wine, a Book of Verse—and Thou
>   Beside me singing in the Wilderness—
> And Wilderness is Paradise enow. [15]

It sounded like an erotic picnic, but bread and wine were religious symbols, the book of verse could be songs such as mystics composed, and they often retreated to the desert. 'Thou' could be a human lover, or the divine, and, as a later verse says, when the veil falls there is no longer any distinction of subject and object, 'no more Thee and Me'. This is only vaguely and symbolically sexual, and it has only a distant relationship, if any, to the sensual embraces of Shiva and Shakti or Yang and Yin.

THE STATUS OF WOMEN

(a) Pre-Islam and Early Islam
Free women in pre-Islamic times and in early Islam seem to

---

[15] *Rubaiyat*, verse 35 in my own version, and verse 11 in Fitzgerald's.

have been a good deal more independent than they were with the veiling and seclusion of later societies. There was a matriarchal family system, in which kinship was reckoned through the female side, and the woman would be controlled by her own family rather than by her husband. Women could choose their husbands and returned to their own family if they were ill-treated. Sometimes they offered themselves in marriage, and Muhammad's first wife Khadija did just this to the Prophet who though of good family was poor and an orphan.

Bedouin poetry sang of the nobility and devotion of women and inspired a 'chivalry' in which horsemen set out to defend their virtue. Stories were told of the courage of women, sometimes defending their menfolk from their enemies at the risk of their own lives. Their physical charms were fully described, with less attention to their moral qualities.

In the Jahiliya, the 'time of ignorance', however, there was a prevalent custom of burying female infants alive, and this was condemned by the Qur'an: 'When the announcement of a female child is made to one of them, his face becomes clouded with darkness and he is full of suppressed emotion. He skulks from the people because of the evil of what has been announced to him'; and again, 'When the female child shall be asked, for what sin she was put to death.' (16.60f.; 81.8f.) Whatever economic arguments there may have been for this practice, they were overthrown by Islam.

In the early days of Islam women had more independence than later. Ayesha refused to veil herself from anyone and said, 'Since God almighty has sealed me with the stamp of beauty, I desire that men should behold it and recognize his grace towards them, and I will not veil it, for there is no defect in me that anyone should speak of it.' Women had a freedom in worship in the earliest days also, and the wife of Umar used to attend morning and evening prayers in the mosque with the men. When she was questioned she retorted that the Prophet had ordered, 'Do not prevent the handmaids of God from access to the places where he is worshipped.'[16]

---

[16] See M. Smith, *Rabi'a the Mystic*, 1928, pp. 120, 124.

Bedouin women had the right to refuse to marry a man who already had a wife, or to leave him if he took another, and this custom remained in early Islam. But with the success of Islam, and the marriage of captive women with their conquerors, it is said that Muslim men preferred slaves as wives because they were not so independent as the free Arab women. The practice of polygamy, temporary marriages, and the seclusion of the harem, with the wearing of the veil, combined to curb the earlier female freedoms.

The restriction of female freedoms was social as much as religious, and obtained particularly in the great Middle Eastern empires, to survive today most strongly in the Arab kingdoms. Medieval Arab travellers remarked on the independence and unveiled state of Berber women, and in some African countries still Muslim women are little or not veiled. Women were active in scholarship, especially in Spain, and many female authorities were preservers of Hadith traditions. Women mystics, too, were notable in a number of countries.[17]

(b) Veiling

With the growth of the Prophet's household his own position was strengthened, and visitors were warned not to enter his houses without warning or stay on after a meal. Similarly they should not ask things from his wives except from behind a curtain, and the wives of the Prophet were told that as they were 'not like ordinary women' they should remain in their houses and not swagger about in the manner of former paganism or wheedle in their speech 'so that one in whose heart is disease grow lustful' (33.32 and 53).

Further Muhammad's wives and daughters, and womenfolk of believers, were told to 'let down some part of their mantles over them', so as not to be insulted but recognized' (33.59). And again:

Say to the believing women that they cast down their eyes and guard their private parts, and show not their ornaments except so far as they normally appear; and let

[17] See 'Women in the Hadith Literature' in *Muslim Studies* by I. Goldziher, tr. S. M. Stern, 1971, Vol. 2, pp. 366 ff.

them throw their scarves over their bosoms, and not show their ornaments except to their husbands or their fathers . . . or male followers who have no desire, or infants who have not got to know about the privy parts of women.' (24.31)

Such instructions would be chiefly for normal covering of the body in public, rather than the free and easy dress of home, and none of the verses clearly indicated veiling the face. Various stories later suggested reasons for these rules, such as that male guests had touched the hands of Muhammad's wives, or that when women had to go out at night for toilet, since there was no indoor sanitation, they were sometimes insulted by unbelievers, deliberately or mistaking them for slaves. Perhaps some men had tried to obtain favours from Muhammad by getting one of his wives to make the request, but since private interviews between men and women who were not related were viewed with suspicion, it was necessary to give the wives some protection, especially since tongues had wagged about Ayesha and the young man in 'the affair of the lie'.

The compulsory veiling of women does not seem to have become established, however, until the centre of the Islamic empire had moved to Baghdad. Even when women had to wear veils in public, the chroniclers do not bother to give details about the masses of the people and in the country women often toiled alongside men. In many middle eastern lands the traditional outdoor garment came to be a large, loose gown from head to foot, with a face-veil (*burka*), and a long strip of white muslin concealing the whole of the face except the eyes and reaching nearly to the feet. But there was great variety, and E. W. Lane, writing of *Modern Egyptians* even in 1836, said that Egyptian women thought it more important to cover the upper and back part of the head than the face, while 'many women of the lower orders, even in the metropolis, never conceal their faces.'

(c) Harems
The veiling of women, and the polygamous household, led not surprisingly to the seclusion of wives and concubines in

private apartments, away from the gaze of strange men. Since the balance of the sexes is usually about equal, except in time of war, if a man has two or more wives, another younger and poorer man may have none and this would give rise to jealousy and intrigue.

The word Harem (*harim*), used in the Middle East, meant 'forbidden' to other men, from an Arabic word *haram* which in another sense, like a similar Hebrew word, meant 'sacred'. In Iran and the Indian sub-continent such female apartments were called *zanana*, from a word for 'woman', and *purdah* from the 'curtain' which screened the occupants from strangers. Scriptural support for such seclusion of women was taken from the verses quoted above, though they may not require more than the covering of female believers' bodies and restricting the Prophet's wives behind a curtain during visits. But it was easy to extend such restrictions, especially in large and rich households, though the poor must have had little privacy.

Descriptions of life in harems were necessarily limited. In general only the husband, among adult men, was allowed to see his wives and female slaves unveiled, though any women, and even non-Muslim women, have been allowed in recent times to enter harems. The wives themselves were not prisoners, for they could go out and pay visits, correctly veiled, and often in curtained litters. Mrs Meer Ali, an English lady who married a Muslim and lived in a Zenana in Lucknow for twelve years, provided a vivid account of life there in the nineteenth century. At first she pitied the monotony of the lives of the women, but thought that never having known liberty they were happy with their confined lot. The women were very fond of company, not being kept from their own sex, and they both visited others and entertained at home. They ate a variety of food, except in the fast of Ramadan, and they smoked hookah pipes and offered them to visitors.

Female slaves abounded in rich harems, but male servants were usually forbidden to the harems of respectable Muslims. Eunuchs, however, were allowed to see the face of any woman, and so were young boys. Eunuchs were often, though not always, black slaves who had been castrated for the service of harems. Stories in the Thousand and One Nights and other romances described their sexual intimacy, though non-procreative, with some of the highest women. Non-Muslim writers expressed shock at the markets which were held for male and female slaves for harems, and regretted that some victims were supplied by Christian traders.

With eunuchs and male slaves homosexual relations flourished, though their frequency is impossible to estimate from the lack of reliable statistics. In addition to innumerable prostitutes, there were said to be many effeminate men who dressed as women, and foreigners often accused Muslims indiscriminately of sodomy. Yet homosexuality was forbidden in the Qur'an and the Traditions.

In modern times there have been many changes, some westernization and some regression to earlier customs. Many Muslim women normally wear western dress for work, in countries like Egypt where perhaps some ninety per cent are unveiled. But in neighbouring Sudan and Arabia practically all women are veiled, except for some very sophisticated ladies or prostitutes. Secularized lands, such as Turkey, have abolished the veil, but in Pakistan young ladies have worn nylon veils which enhance rather than hide their charms. In the struggles of modern Iran educated women have been told by religious enthusaists to cover their arms and faces, but it was said that they would be able to choose their dress 'provided that they observe some guide-lines'. Veiling and harems and clitoridectomy are opposed by reformers and feminists, but supported by traditionalists as part of the defence of Islam against western ways. Tensions between these opposing views are bound to continue for a long time.

Chapter 9

HEBREW AFFIRMATIONS

CREATION

Ancient Hebrew attitudes to life in general and to sex were naturalistic, and religious in the sense of accepting them as divine creations. Judaism was 'natural' and 'classical', in giving due place to human nature. But it always looked beyond man to God in a non-classical way; rather than seeing man as the measure of all things, it portrayed man as he stands before God.

The poetic description of creation in the first chapter of Genesis, a sophisticated account by priestly writers, indicated the stages of divine commands, the appearance of new phenomena and the refrain each time that 'God saw that it was good'. To complete the creation mankind was made in the image of God, 'after our likeness', and in both sexes: 'in the image of God he created him, male and female he created them.' There was hardly a suggestion of dual sexuality in the Godhead, though such a notion was common enough in the ancient world, but the priestly writers were reformers and tried to avoid the polytheism of the past. Nevertheless, male and female were both 'very good', created and ordained by God in the divine image.

The sexual purpose given to male and female was procreation, continuing the divine creation. The first human beings were told to 'be fruitful and multiply', so that sexual intercourse was used in order to perpetuate the race. The world was a good place, to be used and developed, and mankind was told to 'replenish the earth, and subdue it; and have dominion' over all animals of sea, air, and earth. There

178

was no world-renunciation here, and Judaism has generally been opposed to both celibacy and asceticism.

The second account of creation in the succeeding chapters of Genesis, perhaps earlier traditions from so-called Yahwistic writers because of their use of the divine name YHWH, was both cruder and interesting. Instead of orderly stages of creation by simple but powerful divine words, culminating in men and women made in the divine image, here earth and heaven suddenly appeared in a day. Man was moulded like a clay doll, into whose nose God blew 'the breath of life', and woman was made as 'a helper' and companion out of Adam's rib. Adam, 'of the earth', called her Eve, 'life', and she was 'bone of my bone and flesh of my flesh'. They were naked and unashamed.

The story of the 'Fall' by disobedience, the clothing of Adam and Eve, and their expulsion from Eden, has had many interpretations. But the traditional commentators did not interpret the Fall as a demonstration of the temptations or evil of sex. In fact the narrative stated clearly that the sexual union of Adam and Eve came later, after their departure from Eden when 'the world was all before them'. The Fall was disobedience to the divine command not to eat of 'the tree of the knowledge of good and evil'. The 'original sin' was rebellion against God, 'Man's first disobedience, and the fruit of that forbidden tree'.

The story was complicated because a Babylonian myth, which the Garden of Eden tale resembled, was told to account for the existence of death. Apparently, though not clearly, the Genesis narrative held that death would result from eating the fruit, though whether immediate death for Adam and Eve 'in the day that you eat of it', or eventual death for all mankind has been long debated. Milton thought that the tree's 'mortal taste brought death into the world and all our woe.' The punishment awarded for eating the forbidden fruit was not immediate death but sorrow in conception and parturition for the woman, a curse on 'the ground' for Adam's sake, and his own toil and sweat of his brow. Then the couple were expelled from Eden lest they 'take also of the tree of life', apparently another sacred tree, which would make them 'live for ever'.

The talking snake was a curious element in the story, and it was not identified as Satan or the devil. When the snake told the woman that 'you will not die' it may have been because snakes were often regarded as immortal, for they shed their skins but continue to live. The snake is also a common phallic symbol, and in some countries it was believed to have taught the method of sexual intercourse to human beings. In Genesis the snake said that when they ate the fruit, 'your eyes shall be opened, and you shall be as gods, knowing good and evil.' The fruit which Adam and Eve ate was not named, certainly not as an apple, and it might have been a fig or a peach with sexual symbolism. Immediately after eating it Adam and Eve knew that they were naked and made themselves girdles of fig leaves. After their expulsion, Adam 'knew' his wife and went in unto her so that she conceived and bore a son.

The second Genesis story was complex, like most myths, though not half so involved as the Babylonian counterpart. There may have been sexual elements in it, but they were not evil or against the will of God. Common notions that the Garden of Eden story taught an Original Sin which infected all mankind, or even Total Depravity with inherited guilt, have no foundation in the Old Testament, nor was there any suggestion that it was a sexual fall.

A further point to note is the unity of human nature in the biblical teachings. Body and soul were closely bound together, so that it does not say that man was supplied with a soul but he 'became a living soul'. The Hebrew word for 'soul' (*nephesh*) was close to modern ideas of 'personality', and indicated the whole individual. Even after death, regarded as miserable and shadowy, real hope came with the belief in a 'resurrection body'. This psychological unity was very different from the Hellenistic dualism, with its contrast of physical and spiritual, which appeared from the time of Paul and affected later Christian teaching on sex.

PHALLICISM AND CIRCUMCISION

Other phallic elements may be traced in the Bible, though it is a much-expurgated book. Standing rocks and stones regarded as sacred may have been phallic symbols. Moses was said to have had a rod which changed into a serpent, and he made a brazen serpent when the people were bitten by snakes, so that any man 'when he looked at the serpent of brass, he lived.' The people still burnt incense to this snake in the time of king Hezekiah.

The golden calf, or bull-calf, may have been a sexual symbol, since bulls were often credited with great sexual power. When Aaron made the golden calf, 'the people sat down to eat and drink, and rose up to play.' On return from the mountain Moses heard 'the voice of those that sing', and he saw that 'the people were broken loose', and that Aaron 'had let them loose.' It is not hard to imagine an orgy (Exodus 32). When Jeroboam of northern Israel later set up his independent kingdom, golden bull-calves were erected at the shrines of Bethel and Dan, and Jeroboam repeated the words of Aaron, 'This is your god [or gods], O Israel, which brought you up out of the land of Egypt.' Since Yahweh had brought them up he would be represented by a bull; but for the scribes at Jerusalem this was the great crime of Jeroboam 'wherewith he made Israel to sin.'

The clearest phallic element in the Bible was the practice of circumcision. The origins and purpose of this rite are obscure. It is common to most Semitic peoples, as well as Egyptians and many African tribes, and it is practised by Muslims and Christian Ethiopians. But circumcision has been foreign to most European, Indian, Chinese, and neighbouring peoples. There were three accounts of the introduction of circumcision in the Bible. The priestly writer put it back to Abraham, who took it as a sign of a covenant with God, and circumcised himself, his son Ishmael and all his household. The Muslims trace the origin of the Arabs back to Ishmael and Abraham. The Deuteronomic writer made Joshua recircumcise the Israelites 'with knives of flint' when they entered the land of Canaan.

But the most vivid, and probably the oldest, account was given by the Yahwist writers in the story of Moses (Exodus 4). It was said that Moses was returning to Egypt, with his wife Zipporah and their sons on a donkey, and he had the rod of God in his hand. Thereupon God met him at a lodging place and 'sought to kill him', perhaps because he or his son had not been circumcised as was the custom in Egypt. Then Zipporah 'took a flint and cut off the foreskin of her son, and cast it at his feet; and she said, Surely you are a bloody bridegroom to me.' So God left him alone.

The word 'feet' may have been a euphemism for genitals. But whether circumcision was a substitute for human sacrifice has been disputed. The word means something 'cut around' and devoted to God by being destroyed. It was consecrated to the deity in sacrifice, and symbolized the covenant and offering of man to God. For the Hebrews, the uncircumcised man had broken the covenant and would be 'cut off from his people'. Philistines, being Europeans, were uncircumcised and therefore despised, as were similar peoples.

The use of flint knives, in the stories of both Moses and Joshua, showed that the practice of circumcision was very ancient, in vogue before metal knives were known. What the purpose of circumcision was originally is hard to say and various theories have been put forward, of which the most likely may be that it was a preparation for sexual intercourse by removing any tightness of the foreskin. Many tribes to this day perform the rite not upon babies but upon youths, as part of the ceremonies of initiation into manhood. Both Abraham and Joshua were said to have circumcised young men, when the ceremony was supposed to have been introduced. But, rather like Christian baptism later, Jewish circumcision came to be performed on the eighth day after birth and was followed by naming the child.

The story of Zipporah might suggest that circumcision was at first performed by the mother and some writers have suggested, without much evidence, that it was a sacrifice to a goddess of fertility. But the Hebrews did not practise clitoridectomy or female circumcision, which might have been more appropriate for such a goddess. The Abraham story shows that the father came to take the lead, and in later Jewish usage there were special surgeons for the operation. Circumcision was performed in the home, though in the early Middle Ages it was transferred to the synagogue and performed in the presence of the congregation, with appropriate hymns to make it a festal occasion.

Circumcision has been the mark of the Jew down the ages, with few exceptions. In the Hellenistic period Greeks and Romans sneered at the Jews because of their circumcised state, and both Antiochus Epiphanes and Hadrian tried to suppress it. Some Jews who wished to take part in public games and other activities which required nakedness, underwent operations to lengthen their foreskins again, but such practices were strongly opposed by the orthodox. In Jewish reform movements of the nineteenth century circumcision came under criticism, and especially the regulation which required it to be imposed upon proselytes converted to the Jewish faith. Since this necessitated operations on adults it was felt to be too harsh, and in 1892 the American Reform Movement dropped the requirement of circumcision of converts, which was followed by reformed synagogues elsewhere.

Circumcision was the sign of the covenant with God, 'you shall be circumcised in the flesh of your foreskin, and it shall be a token of a covenant between me and you' (Genesis 17, 11). All sons and male slaves in the household were brought within the covenant by circumcision. However, reformers and prophets warned against the formality of external circumcision and turned the language, rather curiously, to internal symbolism. Thus the Deuteronomist exhorted the Jews to 'circumcise the foreskin of your heart', and Jeremiah accused them of being uncircumcised in both heart and ears.

Circumcision was a problem for the early Christians, who were divided between the Judaizers who said, 'Unless you are circumcised after the custom of Moses you cannot be saved', and those like Paul who, despite his circumcising of Timothy, decided that 'we trouble not those among the Gentiles who turn to God' (Acts 15). Paul came to be entrusted with 'the gospel of uncircumcision', as Peter was with 'the gospel of circumcision'. When the Christian Church became almost entirely Gentile circumcision was abandoned, except for some sects like the Judaistic Ebionites. Verses concerning spiritual circumcision, initiated by the prophets, remained in Christian writings though they may sound strange to modern congregations, if they listen:

'We are the circumcision, who worship by the spirit of God', and 'in him you were circumcised with a circumcision not made with hands.'

MALE AND FEMALE

The Bible had a strong patriarchal emphasis, from Abraham, Isaac, and Jacob, with their plurality of wives, and this helped male dominance. Legally women were in a position of inferiority and the wife was 'possessed' by her husband. In the development of the Law both sexes had religious responsibilities, though women were exempted from some requirements, such as the wearing of phylacteries. Man was chiefly responsible for carrying out the precepts of the Torah, whereas woman had domestic obligations. The Talmud prescribed a thanksgiving by men that they had not been made women, and this is still recited from the authorized Jewish Daily Prayer Book: 'Blessed art thou, O Lord our God, King of the universe, who hast not made me a woman.' Women for their part bless God 'who hast made me according to thy will'.

The Talmud often spoke in a disparaging way of women, but not of her position in the home where she had a vital religious function. Every Sabbath the mother of the house, surrounded by her husband and children, kindles the candles with a blessing to God 'who hast sanctified us by thy commandments and commanded us to kindle the Sabbath light'. The father then takes a cup of wine and recites over it the Kiddush or hallowing of the day, sips from the cup and passes it to the wife and children. It has become the practice also for the father to recite the praise of the virtuous wife from Proverbs which follows.

The virtuous woman, whose 'price is far above rubies' is described in detail and glowing terms in Proverbs 31. She gets up before dawn to feed her household, she works wool and flax with her hands, and spins with a distaff and spindle. She buys a field, plants a vineyard, gives food to the poor and needy, clothes her household with scarlet, and makes linen garments and girdles for sale to the merchants. Her children call her blessed and her husband praises her. But what is he doing all this time? He is sitting at the village gate with the elders of the land, perhaps deciding legal cases or perhaps gossiping.

There were similar, but more ambiguous, words in the apocryphal Ecclesiasticus (25–6). 'Happy is the husband of a good wife', a brave woman, a shamefast woman, a silent woman; and 'the beauty of a good wife' is like the sun when it rises in the high places. On the other hand, give me 'any wickedness but the wickedness of a woman', which changes her looks, darkens her face like a bear, and makes a lion or a dragon preferable. A wife full of words is like a sandy path to old people, a wicked woman is like a shaking yoke of oxen, a drunken woman causes great wrath, and the whoredom of a woman is in the lifting up of her eyes. 'From a woman was the beginning of sin, and because of her we all die.'

Such misogyny was echoed in the Talmud which criticized the garrulity and jealousy of women, but whereas some verses said that women were light-minded and cared only for beauty, another verse admitted that 'God endowed women with more intelligence than men.' As in some other cultures, Jewish women were often said to be more addicted to witch-craft and occult practices than men. The so-called 'witch' of Endor (called a 'witch' only in the headings of the Authorized Version) was clearly a spiritualistic medium and she was said to have had a 'familiar spirit'. That women are often mediums in many countries helps to explain the feeling that their sex was particularly addicted to the occult, and men were the guardians of more orthodox and public religious ceremonies.

The patriarchal emphasis of the Old Testament affected concepts of God, who was regarded as a patriarchal ruler of his people, and a stern monarch. God was said to be almighty, all-knowing, all-seeing, jealous and death-devoting of his enemies. It is curious that these all too human feelings did not include sexual activity, in the manner of the Indian Shiva. But the stern decrees and nationalistic sentiments expressed in the Old Testament have often offended people in more humane times, to the neglect of the more gracious features of the picture. Modern critical study of the

Bible has made much of the concept of development, from the primitive to the reforming and prophetic, and early beliefs can be seen better in the context of their times.

In sexual matters harsh penalties were decreed against those who deviated from the 'normal' man–woman relationship which should culminate in marriage. Practices that were considered to be against nature as designed by God, were commanded to be given stern condemnation and punishment. Loose women were dangerous. The foolish woman was clamorous, and called to passers-by to enter her house with the promise that 'stolen waters are sweet', but the unwary man did not know 'that the dead are there'. (Proverbs 9)

The priestly and holiness code of Leviticus decreed the death penalty for male homosexuality: 'If a man lie with mankind, as with womankind, both of them have committed abomination: they shall surely be put to death; their blood shall be upon them.' (20.13) There was no mention of Lesbianism here, though it was later referred to by Paul in the Hellenistic world: 'women changed the natural use into that which is against nature', so that 'the wrath of God' was revealed against them (Romans 1.26). But there is no foundation for the notion that Sodom and Gomorrah were destroyed for homosexual practices, and the Bible and apocrypha describe their sin as 'pride', 'prosperous ease', and neglect of 'the poor and needy' (Ezekiel 16.49). The notion of 'sodomy' as homosexuality seems to have appeared among Palestinian rigorists who hated the Greek way of life and attributed such practices to it. Traces of the later belief may be in New Testament reference to Sodom and Gomorrah having 'given themselves over to fornication' and gone 'after strange flesh', and this influenced later Christian thought. (Jude 7)

Transvestism was condemned in Deuteronomy (22.5), both of a woman wearing a man's clothes, or a man putting on 'a woman's garment', which was said to be an abomination to the Lord. Bestiality was strongly condemned in Leviticus, if a man or a woman lie with a beast 'they shall surely be put to death,' and 'you shall slay the beast' (20.15–16).

Two long lists (Leviticus 18 and 20) give prohibited degrees of affinity, which probably went back to pre-Mosaic times and formed the basis of later Jewish and Christian marital prohibitions. The curious phrase was used 'to uncover the nakedness' of any that was near of kin, indicating the most direct intercourse. Although intercourse with a brother's wife was forbidden, a later text commanded a brother to take his deceased brother's wife, if she had no son, and 'raise up seed to his brother', and this practice continued into New Testament times (Deut. 25.5). This Levirate marriage (*levir* meaning brother-in-law), implied that the first child would be named after the dead brother, and the story of Onan (see below) showed that some men disliked the idea. Deuteronomy said that if the man did not want to take his brother's widow he should tell the elders, and then the woman would take off his shoe and spit in his face, and curse him as one who did not build up his brother's house. The Talmud commented that the elders would advise the man against the marriage if he was young and she old, or she young and he old, for that would introduce strife into the home, and a man should marry a woman about his own age.

Leviticus prescribed Draconian punishments for adultery, when 'the adulterer and the adulteress shall surely be put to death'. A similar penalty would follow intercourse with a father's wife or a daughter-in-law. If a man married a woman and her mother they should be 'burnt with fire, both he and they'. To lie with an uncle's wife was supposed to entail childlessness, as with a brother's wife. Sexual intercourse during menstruation would entail expulsion from the community, for 'he has made naked her fountain, and she has uncovered the fountain of her blood.'

Virginity was treasured and the Deuteronomist gave details to attempt to safeguard it. If a man married a wife and consummated the union but disliked her, he might claim that she was not a virgin. Then her parents would produce 'the tokens of virginity', the blood-stained nuptial cloth, and 'spread the garment before the elders of the city'. If this was accepted as proof of the ruptured hymen the man should be fined for dishonouring the virgin's name and he would be bound to her for life. But if the tokens of virginity were not found the girl should be stoned to death, 'because she wrought folly in Israel, to play the harlot in her father's house.'

If a betrothed virgin lay with another man in the city, they should both be stoned to death, 'because she did not cry out, being in the city.' But if he raped her in the field, then the man alone should die, for 'she cried and there was none to save her.' Yet if the virgin was not betrothed and 'a man should lay hold on her and lie with her,' then he should pay a fine and she should become his wife for life, 'because he has humbled her'. (Deut. 22)

Blood and semen both brought ritual impurity. Menstruation and any other female flow of blood required purification, with washing of clothing and of anyone who had touched the woman. After childbirth the mother was considered unclean for seven days until the male child's circumcision, or two weeks for a girl, and there were thirty-three days of purification, during which time she could 'touch no hallowed thing nor come into the sanctuary.' Finally, she had to take a lamb or a dove for 'a sin offering' to the priest who would 'make atonement for her' to the Lord. (Lev. 12) Blood was very powerful and there were ancient taboos against its polluting energies. In another context, the blood was regarded as the life or the very soul of a living being, hence all the *kosher* regulations against eating the blood of an animal.

Semen too brought uncleanness, in copulation, or involuntary emission, or continual emission. The man would be unclean until the evening and all his clothes needed washing. The woman also with whom he lay would need to bathe in water and remained unclean until the evening. A man with mutilated penis was forbidden entry into the 'assembly of the Lord', and a bastard also was banned and his children 'even to the tenth generation' (Deut. 23), a regulation that must have been hard to enforce but which demonstrated the priestly concern for ritual purity and regularity in the community of the Lord.

The story of Onan (Gen. 38), who was ordered to have intercourse with his dead brother's wife but at the moment of coition spilt his seed on the ground lest he should give seed to his brother, stated that he was killed by the Lord for this evil action. The *Oxford English Dictionary* defines onanism as 'uncompleted coition' or 'masturbation' or 'self-abuse', but it was certainly not the last two. The large dictionary quotes a 'medical' opinion of 1874 that 'Onanism is a frequent accompaniment of insanity and sometimes causes it', and no doubt popularization of this notion caused needless anxiety to masturbating boys in Victorian and later Britain and America, as witnessed by *Portnoy's Complaint*.

The woman in the Onan story, Tamar, was determined to get impregnated; she put off her widow's clothes, disguised herself as a prostitute and seduced her father-in-law, Judah. She cleverly kept as pledge his signet and cord and staff, and when she was accused before Judah of harlotry and condemned to be burnt, she produced the pledges and was cleared. She bore twins from this finally consummated family relationship, and Tamar was quoted in Matthew's Gospel as an ancestor of Jesus.

LOVE AND MARRIAGE

It is remarkable that the Bible which, although it was world-affirming, gave strict and masculine regulations about the relations of the sexes, should contain such a rapturous and sensual book as the Song of Songs. It might seem to be more suited to the environment of Hinduism, among the love lyrics of Krishna and Radha, and it has been compared to the Gita Govinda. Yet the Bible is an anthology, of many dates and by countless hands, and with very diverse subjects. The Song of Songs is not religious and it never mentions the name of Yahweh, yet it has given pleasure to Jews and Christians down the ages.

The place of the Song of Songs in the canon of Jewish scripture was disputed 'until the men of the Great Assembly came and expounded it' in a spiritual sense. Rabbi Akiba said that 'the whole world was not worthy of the day when it was given to Israel. All the Scriptures are holy, but the Song of Songs is holy of holies.' The debate over the book was whether it was a love idyll or an allegory of the relationship between God and Israel, representing in symbolical language the divine-human communion. The latter view prevailed, and the book was said to have been written by Solomon and inspired by the divine Spirit.

In the Christian Church this allegorical interpretation went much farther. Bernard of Clairvaux in the twelfth century expounded the book as teaching the mystical union of God and man, in his sermons on the Canticles. This application was taken over in editions of the English Authorized Version of the Bible, where the page and chapter headings explained as 'a description of Christ by his graces' the sensual picture of the lover with 'his lips like lilies . . . his belly as bright ivory'. And 'the graces of the church' were the explanation of the picture of the beloved with 'lips like a thread of scarlet' and 'two breasts like two young roes'.

Both Solomonic authorship and the allegory of divine love are unnecessary for the appreciation of this work as a collection of love lyrics. It may have been a drama of the infatuation of a chief for a shepherd maiden who already loved another, or a marriage play celebrating true betrothed love. Most likely it was a series of poems of love which were sung or recited at wedding feasts in ancient times. Similar poems are still chanted at marriage festivals in the Near East, where bridegroom and bride are addressed as king and queen, as they are in India. The 'king's week' among peasants enlightened the hard round of agricultural life, and the love of man and woman was associated with the joys of spring, as in old fertility rites:

> Rise up, my love, my fair one, and come away.
> For, lo, the winter is past,
> The rain is over and gone;
> The flowers appear on the earth;
> The time of the singing of birds is come.

Much of the poem is in the woman's voice, showing her joy in sex: 'Let him kiss me with the kisses of his mouth: for thy love is better than wine.' She feels concern for her dark skin but evident beauty: 'I am black, but comely.' She sees the king at his table but awaits her beloved, 'leaping upon the mountains'. At night she seeks him in the city streets (in a 'dark night of the soul' for mystics): 'In the streets and in the

broad ways, I will seek him whom my soul loves; I sought
him but found him not.'

Then the lover takes over and praises all the parts of her
lovely body: 'You are all fair, my love, and there is no spot in
you.' Again she sleeps and hears him at the door, but when
she arises he is gone, and she is left to justify her love by
describing his beauty in turn: 'My beloved is white and
ruddy, the chiefest among ten thousand.' Later there come
further female descriptions, this time from the feet up: feet
in sandals, joints of thighs like jewels, navel (or vulva?) like
a goblet with mingled wine, belly like a heap of wheat,
breasts like fawns, neck like a tower of ivory, eyes as dark
pools, nose like a tower, and hair like purple to hold the king
captive in its tresses.

No paraphrase can do justice to the beauty of these verses
which celebrate the love of man and woman in inimitable
terms and conclude,

> Set me as a seal upon thy heart,
>   as a seal upon thine arm,
> For love is strong as death.

Hebrew marriage was a religious duty, and one rabbi said
that a man who does not marry is not fully a man. Another
rabbinic saying was that 'every man needs a woman and
every woman needs a man, and both of them need the divine
presence.' Other sayings were that 'a man's home is his
wife', and 'I have never called my wife by that name, but
always "my home".'

Judaism, like Islam, deprecated celibacy. Priests and
rabbis should be married, and the High Priest was com-
pelled to marry. There were a few exceptional celibates and
Jeremiah was told, 'You shall not take a wife, nor have sons
and daughters in this place.' (Jer. 16.2) It is not clear whether
this was to be a permanent state, since this word came to
Jeremiah in Jerusalem, but afterwards he went to Egypt.
Even so, it was a particular command for special reasons,
and other prophets were married. Apparently the later
Essenes thought that there was some religious value in occasional or regular abstinence from sexual intercourse, but
this was against the tradition of orthodox Judaism, though it

may have affected early Christianity.

The term for the marriage ceremony was *Kiddushin*, meaning 'sanctification', related to the general word for 'holy'. Marriage was both a sanctified relation and one in which, according to the Talmud, 'the husband prohibits his wife to the whole world, like an object which is dedicated to the Sanctuary.' It has been denied that Hebrew marriage was a sacrament, and it did not have the priestly accompaniments of later Christian marriage. But perhaps it was nearer to the simpler sacramental concept of husband and wife in a holy relationship which came from the spirit that they brought to it, and in which they ministered to each other their vows before God. The Talmud suggested that since the creation down to the present, God has been arranging marriages, and since marriages were thought to be made in heaven there is some ground for the opinion that the romantic view of marriage is the religious view. At the wedding of Adam and Eve, God acted as groomsman for Adam and plaited Eve's hair to adorn her for her husband.

Men were recommended to marry from the age of eighteen; traditionally building a house, planting a vineyard, and taking a wife. It was said that 'up to the age of twenty, the Holy One, blessed be he, watches for a man to marry, and curses him if he fails to do so.' A girl should be married even earlier, having ceased to be a minor at the age of twelve, and being given the choice to say, 'I wish to marry so-and-so.' The importance of pleasure in marriage and procreation was emphasized by the regulation (Deut. 24.5) that 'when a man takes a new wife he shall not go out with the army, nor shall he be charged with any business. He shall be free at home one year, and shall enjoy the wife whom he has taken.'

Although it was the father's duty to obtain a husband for his daughter, there were guidelines on the appropriate choice. A man who gave his daughter to an old man could be confronted with the Levitical verse which said, 'Do not profane your daughter, to make her a harlot.' Similarly, a young son should not be married to an old woman. Tall, short, fair, or dark men should not marry women of the same kind, lest their characteristics be excessively reproduced in their children.

It is well known that polygamy was practised in biblical times, and it continued to some degree long after. Solomon, wisest of men, was said to have had seven hundred wives and three hundred concubines, but 'his wives turned away his heart' from God because they brought cults of other gods with them. David, on whom the Spirit of the Lord had come, had numerous wives and concubines, but he lusted after Bath-sheba and sent her husband to his death so that he could keep her. Priests as well as rulers could have more than one wife in biblical times, but while polygamy has been prevalent in many eastern countries it is doubtful whether ordinary people practised it extensively.

The model given by the second creation story in Genesis should strictly apply only to a monogamous marriage: 'Therefore a man shall leave his father and his mother, and shall cleave unto his wife; and they shall be one flesh.' Similarly, the picture of the ideal woman in Proverbs seems to have implied monogamy, as do other stories, and the role played by the mother in the Jewish home reinforces this state. One should beware of judging the whole of society from the example of its rich or powerful leaders whose activities have been recorded in much more detail than those of ordinary couples.

Both a cause and the problems of polygamy were illustrated in the narrative of Samuel. His father Elkanah had two wives, but Hannah had no children, which was perhaps why the other wife had been taken. The second wife was called a 'rival', and she 'provoked' Hannah every year sorely, and made her fret and not eat, 'because the Lord had shut up her womb'. Finally, after she had prayed for years, 'the Lord remembered her', and she bore a son and called him Samu-el, 'heard of God', and dedicated him to the divine service.

The Talmud also sanctioned polygamy, though different views were expressed. One authority said that 'a man may marry as many wives as he pleases', but another that 'he may not exceed four' (like the Muslims), while yet another rabbi declared that a man must give his wife a divorce if she wished for it when he took another wife. Polygamy was only finally outlawed in Judaism from the eleventh century of the present era, and excommunication is now imposed upon those who practise it except for a small number of Jewish communities in Muslim countries.

Divorce was easy, for the man. The patriarchal system of the Bible and the Talmud gave absolute authority to the husband, and it was declared that while 'a woman may be divorced with or without her consent, a man can only be divorced with his consent.' But eventually the reformer Rabbi Gershom in the eleventh century, who forbade polygamy, also decreed that no divorce could be effected without the consent of the wife.

The ease of divorce, giving the woman a bill of divorcement because of some unseemly thing, was decreed by Deuteronomy (24) and interpreted by the rabbis. The 'unseemly thing', literally the 'nakedness of a thing', was taken by the school of Shammai to mean that a wife should not be divorced unless she was unfaithful. But the school of Hillel said that it could be anything unseemly, even bad cooking, and this lenient opinion tended to prevail. The words of the prophet Malachi (2.15), 'let none deal treacherously against the wife of his youth, for I hate putting away, saith the Lord', was interpreted as meaning, 'if you hate your wife, put her away.'

A safeguard against hasty divorce was that it entailed repayment to the wife of the marriage settlement (*Kethubah*) and this could be a crippling burden for a man who had a difficult wife but would have to pay a large sum to her if they parted. However if her conduct caused scandal she could be divorced without receiving the *Kethubah*, and bad conduct could be taken in various ways: making loud cries which could be heard in public, cursing the husband's children in his presence, talking with all sorts of men, or even going out with uncovered head. Insanity was not a ground for divorce, lest an outcaste woman become the prey of an evil-minded person, but serious disease such as leprosy would justify divorce.

That marriage could be ended easily, especially by the man, was said to have helped to maintain the standard of most marriages that remained. Moreover, the importance of children, especially sons, made most people persevere in their marital relationships. A large family was a sign of divine favour, and it brought with it the corresponding obligation to bring up the children with the greatest care.

SYMBOLISM

A patriarchal system had a patriarchal God, and the monotheism developed by the Hebrews, with its ethical concomitants, was their most notable achievement. It was more clear cut than Hindu monotheisms, and it inspired Christianity and Islam. It might seem that there could be no sexual symbolism in concepts of such a transcendent deity, yet at early and late periods there were signs of diversity within the unity.

At first there was the great struggle with polytheism, and female and fertility deities. Solomon's wives turned away his heart, notably towards Ashtoreth or Ishtar the goddess of Sidon, and apparently a temple for this goddess was at Jerusalem until it was defiled three centuries later by Josiah. Ahab, King of Israel, married Jezebel of Sidon and she brought foreign cults with her, probably including female divinities. Israel fell in the eighth century, and the monotheistic reforms of prophets and priests were most successful in the small kingdom of Judah around Jerusalem, which also fell in the sixth century.

When Jeremiah was carried off to Egypt he was shocked to find Jewish women there worshipping Ishtar, the Queen of Heaven, with cakes and drink offerings. These were justified by their men, saying that they would 'burn incense to the queen of heaven and pour out drink offerings to her, as we have done, and our fathers, our kings and princes, in the cities of Judah and in the streets of Jerusalem.' (Jer. 44.17) Jeremiah prophesied destruction for them, but that they were not alone in worshipping a goddess has been shown from the papyri of Elephantiné, an island in the Nile in Upper Egypt, opposite Aswan. These writings which belonged to a Jewish military colony, founded at an uncertain date, mention the worship of Yahweh but also of other gods of whom one, Anathyahu, bore the name of the female deity Anath combined with Yahu, and this suggests that she was regarded as the spouse of Yahweh.

The relations of God and man involved duality, which could be expressed by sexual symbolism. Occasionally feminine language could be applied to God, but only metaphorically: 'Can a woman forget her sucking child? . . . yet I will not forget you', and again, 'As one whom his mother comforts, so I will comfort you.' (Isaiah 49.15; 66.13) But normally God was the male, father or husband, and the feminine figure was fixed as the human partner.

Sexual language was harnessed by the prophecy of Hosea, in the northern kingdom of Israel in the eighth century. Hosea enacted a parable in his own marriage, being told by God to marry a prostitute, because 'the land plays the harlot by infidelity to the Lord.' The woman, Gomer, bore three children, who were given names indicating punishment and rejection, 'not my children'. Gomer continued in fornication, 'for she said, I will go after my lovers that give me my bread and water, my wool and flax, my oil and drink.' This was a parable of Israel, that had gone after local fertility gods (Baalim), propitiating them to obtain the produce of the land and the flocks, but she did not recognize that these all came from the one national God: 'She did not know that I gave her the corn and the wine and the oil, and multiplied to her silver and gold which they used for Baal.' (Hosea 1–3)

Eventually the faithless woman was called back to her forgiving husband, God: 'She will call him "My husband" (Ishi) and not "My Lord"' (Baali). She will be betrothed forever in faithful wedlock; God will command the heavens and the earth, and there will be peace and prosperity. God will say, 'You are my people', and Israel will answer, 'You are my God.' These themes were pursued throughout the book of Hosea, calling the people from adultery, forgiving them, loving them freely, and drawing them with bonds of love.

A similar model of the divine marriage of God and the people was given by Jeremiah, who accused Israel and Judah of playing the harlot with many lovers. Yet God called the people to return, 'for I am a husband to you', and he would forgive the treacherous wife if she repented. Later, speaking of the covenant, Jeremiah said that God had been 'a husband to them'. (Jer. 3 and 31) Later passages of Isaiah used comparable terms: 'As the bridegroom rejoices over the bride, so shall your God rejoice over you.' (Isa. 62.5)

The transcendence of God was not therefore as stark and harsh as some would maintain, and beside the warlike or jealous God, there was the deity of compassion and nearness. The Bible represented different viewpoints, wherein the cruelty of a Jehu could be censured by a Hosea. In the later Old Testament and Apocrypha there were further developments.

The book of Proverbs presented a series of moral addresses on the value of wisdom, coming to a climax in a beautiful picture of Wisdom personified as the female companion of God before creation. 'The Lord possessed me in the beginning of his way, before his works of old . . . I was beside him like a master workman, and I was daily his delight' (Prov. 8.22ff.) This was an advance on the picture of Wisdom given in Job as 'above rubies' and known only to God (Job 28), and the influence of Greek philosophy has been suggested. Yet Wisdom, for Proverbs, was not eternal but was formed by God as an instrument of creation. This view was developed in Ecclesiasticus (24) where Wisdom said, 'I came forth from the mouth of the Most High, and covered the earth as a mist': Wisdom was established in Jerusalem and was to be found in the Law. In the book of Wisdom, she was said to rise from the power of God, 'a pure effluence of the glory of the Almighty, a brightness streaming from everlasting light, and a flawless mirror of the active power of God.' She is 'fairer than the sun', 'she, being one, has power to do all things', and 'makes men friends of God' (7.25ff).

Later interpretations of these ideas were various. Rabbis taught that the Law was created before the world, and Philo conceived of the Logos, the divine word or power, as immanent in all things and also as coming from God in acts of creation. The suggestion of some kind of associate or power with God, which could be personalized, no doubt helped in the formation of Christian doctrines of the nature of Christ and the Trinity.

For Judaism, it is perhaps enough to note the female nature of Wisdom as described by such writers. In Proverbs it was said that, 'Wisdom has built her house, she has hewn out her seven pillars', she sent her maidens to call men to eat of her bread and wine, and this picture was contrasted with that of the foolish woman who sat at the door of her house calling passers-by to eat secret bread and drink stolen waters. In Ecclesiasticus, Wisdom was said to give off the scent of perfumes and flowers, and again she offered food and drink. While in the Wisdom attributed to Solomon it was said that, 'I loved Wisdom; I sought her out from my youth, and longed to win her as my bride, and I fell in love with her beauty.'

In one tractate of the Talmud such words about Wisdom were applied to the Torah, the Law. It was the Torah whom 'the Lord possessed in the beginning of his way', and without it heaven and earth could not endure. The Torah was pre-existent, preceding the creation by thousands of years, and dearer to God than anything else he had made. The Torah was perfect and could not be improved upon, though despite many analogies it was generally spoken of in a neutral fashion.

In some mystical works, however, the Torah–Wisdom was female, and the Zohar compared her to a stately maiden, secluded in an isolated chamber of a palace, with a secret lover whom she alone knew. He hovered about the palace, trying to get a glimpse of her and she opened a small door for a moment to reveal her face to her lover, and then quickly withdrew it. The lover knew that this momentary revelation was done from love, and so his heart and soul were drawn to the study of the Torah. At first she spoke to him from behind a veil to suit his manner of understanding, then she uttered riddles and allegories in a filmy veil of finer mesh. Finally the maiden was disclosed face to face and conversed with him of all her secret mysteries.

The medieval Jewish Cabbalists, especially the Spanish, did not normally interpret the Song of Songs of the love and union of God and the soul, in the manner of Christian mystics. The only place where the Zohar used sexual language of the relations of a mortal to the divinity were in reference to the Shekhinah. The Shekhinah, 'indwelling', was a Talmudic term for the immanence and omnipresence of God. Moses was believed to have ceased sexual relations with his wife after he had been with God face to face on Mount Sinai, and the Zohar said that Moses had intercourse with the Shekhinah like a mystical marriage.

Sexual language was used however by Cabbalistic mystics in describing the relation of God himself to the Shekhinah. In this mystical mythology God was En Sof, the 'endless', the absolute infinite. From this infinity emanated ten Sefiroth, abstract entities which irradiated the universe. The last and tenth of these, which represented the harmony of all the Sefiroth and the presence of God in the universe, was the indwelling Shekhinah. Owing to the sinfulness of man the Shekhinah was in exile and was only found in isolated individuals and communities. The reunion of the Shekhinah with En Sof was therefore the interest of the Cabbalists, who did not hesitate to use sexual imagery for this object. The mystery of sex was a symbol of the love of God for his Shekhinah, the sacred union of King and Queen, or the Celestial Bridegroom and Celestial Bride.

Many variations of sexual imagery occurred in speculations about primeval procreation, the ray which was sown into the 'celestial mother', the divine intellect from whose womb came the Sefiroth. Phallic symbolism appeared in speculations about the ninth of the Sefiroth, Yesod, 'Foundation', from which all the higher Sefiroth flowed into the Shekhinah as the procreative life force of the universe. The sign of circumcision was taken to show that mystical procreation had its rightful place. [1]

The use of sexual terminology should hardly have been strange in a religion which had a positive attitude to material life, but its application to God himself was a new departure. Yet this could lead to a reaffirmation of marriage as one of the most sacred mysteries, since a true marriage was a symbolical realization of the union of God and the Shekhinah. The phrase from Genesis that 'Adam knew his wife Eve' was taken as indicating that 'knowledge' meant the

---

[1] G. G. Scholem, *Major Trends in Jewish Mysticism*, 1955, pp. 225 ff.

realization of union, like that of the King and the Shekhinah.

In medieval German Hasidism the relations of God and man were spoken of as passionate love. 'The soul is full of love of God and bound with ropes of love . . . The flame of heartfelt love bursts in it and the exultation of innermost joy fills the heart.' Further, this relation was described in terms of erotic passion. Earthly love, which was described in detail, was taken as an allegory of the heavenly passion, drawing 'an inference from the nature of the sensual to that of the spiritual passion; if the force of sensual love is so great, how great must be the passion with which a man loves God.'[2]

Erotic terminology was used in some Hasidic writings of violent movements in prayer which was described as 'copulation' with the Shekhinah. The verse of Job, 'From my own flesh I behold God', was used to express the analogy of a child being born as the result of physical copulation with a 'vitalized organ', with the spiritual copulation, the study of the Torah and prayer, 'performed with a vitalized organ and with joy and delight'. Another statement was even more direct, saying that 'as there is swaying when copulation begins' so a man must sway at first in prayer and then become immobile and tightly attached to the Shekhinah. By swaying a man would attain a powerful state of arousal, because the Shekhinah stood over him and he would attain to a stage of great enthusiasm.[3]

Such language shocked the orthodox and the 'enlightened' opponents of Hasidic enthusiasm, and there was a reaction among the Hasidim themselves towards serving God with the soul alone, and without any bodily movements, though swaying in prayer is practised to this day in some quarters. Women also would be excluded from erotic Hasidic doctrines of prayer as the intercourse of the male worshipper with the female Shekhinah, and while women offered their prayers they were not discussed in the detailed life of prayer described in the classical Hasidic writings.

More generally the Sabbath was spoken of with sexual symbolism. The Sabbath was endowed with divine beauty

[2] Scholem, *Major Trends in Jewish Mysticism*, pp. 95 f.
[3] L. Jacobs, *Hasidic Prayer*, 1972, pp. 60 f.

and became the Bride which was sought with love. At sunset
every Friday evening the Lover, Israel, goes forth to meet the
Bride, the Sabbath, with songs of welcome and praise. A
mystical author, Solomon Alkabetz, used to go out to the
fields with his friends at sunset on Friday to greet the Sabbath bride, and his hymn now occupies an honoured place in
Synagogue rituals:

> Come forth my friend, the Bride to meet,
> Come, O my friend, the Sabbath greet.[4]

---

[4] I. Epstein, *Judaism*, 1959, p. 248.

Chapter 10

CHRISTIAN DIVERSITY

Every religion has some distinctive characteristics and Christianity is the only major religion which from the outset has seemed to insist upon monogamy. Of course Christianity was a religion of reform, developing from Hebrew naturalism, but Buddhism was a reform of Hinduism yet it did not make a similar insistence upon monogamy. In theory monogamy should have offered the best opportunity for equal rights to husband and wife, and the highest regard for married love, but unhappily for many centuries such ideals were not the most cherished and only in modern times are their implications being more widely realized.

THE GENTILE BACKGROUND

Christianity is a complex religion, largely a synthesis of Hebrew and Greek thought, with other influences. It had a background of Hebrew naturalism, in which sex was regarded as created by God, with emphasis on procreation and male-domination. There were ascetic tendencies, notably among the Essenes, but anti-sex drives came from outside Judaism upon the young Christian churches. Christianity moved rapidly into the Greek and Roman worlds, and from being a small Jewish sect it became an international religion; but there were losses as well as gains in the process. From an early period various forms of asceticism affected the new religion: Jewish, Greek, Gnostic, Manichee, and perhaps Buddhist and Jain.

The Greeks were polytheists and their pantheon resembled

that of the Indian Aryans, to whom they were distantly related. But the Vedas were more religious than Homer and the sexual exploits of the Indian gods tended to create more elaborate religious mythology than that of the Greeks. Barnabas and Paul were called Zeus and Hermes at Lystra, at Ephesus the worshippers of Artemis raged against Paul, and at Athens he stood on the Areopagus and may have looked up to the Parthenon of Athene when he preached.

Zeus Pater, like the Indian Dyaus Pitar and Roman Jupiter, was the sky and rain god, protector of the family and customary laws. But mythology attached to Zeus countless amours, and his seductions of Europa, Danae, and Io were popular in art, and especially that of Leda whose rape by Zeus in the form of a swan has continued to fascinate European artists with its phallic symbolism. At the fourth-century Roman villa at Lullingstone in Kent there may still be seen a mosaic in the dining-room depicting Zeus as a bull carrying off Europa, while in a lower room nearby was a small Christian chapel. The very success of Greek art and literature, which grossly personalized the adventures of Zeus may have led to his downfall. For, as Gilbert Murray remarked, the worship of phallic emblems of fertility, which was common throughout the eastern Mediterranean and far beyond, was in itself intelligible and not necessarily degrading. 'But when those emblems are somehow humanized, and the result is an anthropomorphic god of enormous procreative power and innumerable amours, a religion so modified has received a death-blow . . . The unfortunate Olympians, whose system really aimed at purer morals and condemned polygamy and polyandry, are left with a crowd of consorts that would put Solomon to shame.'[1] The Indian cults of the phallic Shiva, however, have continued to prosper.

Criticism of the fables of Homer and Hesiod and other Greek poets came from the philosophers, and Plato in his *Republic* attacked bad representations of the gods and heroes. But his censures were chiefly on stories of the fights and intrigues of the gods, and their breaking of oaths and

[1] G. Murray, *Five Stages of Greek Religion*, 1925, p. 91.

promises, rather than on their sexual behaviour. In his *Symposium* Plato spoke of earthly and heavenly love, the former being the work of the 'earthly Aphrodite', by which men were 'as much attracted by women as by boys'. The heavenly love Plato saw as springing 'from a goddess whose attributes have nothing of the female, but are altogether male, and who is also the elder of the two, and innocent of any hint of lewdness.'

In Greece male homosexual practices were often idealized, and to some degree institutionalized, and they infiltrated later into Roman society. There were also many courtesans, hetairae, and lower grades of prostitutes. The pleasures of the flesh were accepted, though philosophers like Aristotle taught the importance of moderation, *sophrosune*, which indicated a mean between gross sensuality and asceticism, much broader than the Buddhist Middle Way.

Plato was partly responsible for the notion of a dualism of body and soul, an opposition between them which was quite unhebrew and came to have a devastating effect upon Christian views of marriage. Plato wrote of those who called the body the tomb of the soul, as if he was not quite sure about it, but elsewhere he said that the body was a hindrance to the soul, its pleasures were slavish, and it was a source of evil. 'The body fills us with loves and desires and fears and all sorts of fancies and a great deal of nonsense.' So that the man who pursues truth does so by 'cutting himself off as much as possible from his eyes and ears and virtually all the rest of his body, as an impediment which by its presence prevents the soul from attaining to truth and clear thinking.'[2]

In reaction against sexual laxity in classical times, later Greek philosophers often proclaimed an asceticism which involved mortification of the flesh. Their pessimism or 'failure of nerve' tended to abandon the material world, and this was helped on by the low contemporary view of woman and marriage. Although in the fourth century B.C. Diogenes, founder of the Cynics, had claimed that what is natural cannot be indecent or dishonourable and should be done in public, he himself lived in extreme poverty and his followers

[2] *Phaedo*, 66.

repudiated worldly comforts, including marriage and the family. The Stoics sought to live in harmony with reason or nature, which was the only good and everything else was indifferent, but this tended to the rejection of family ties. Even the Epicureans, who sought the 'sweetness of life' did not do so directly but aimed at escape from the world. Epicurus lived in a retreat called the Garden where men and women, slaves and hetairae, found a refuge in a simple life where they took neither flesh nor wine. Their enemies accused them of immorality but Epicurus said that 'the wise man will not fall in love,' and the 'physical union of the sexes never did good; it is much if it does not do harm.' The Neo-Pythagoreans also inclined towards a dualism which involved continence and regarded sexual intercourse as defilement.

Ancient Rome had a higher ideal of marriage than was common among the Greeks, and its primary purpose was procreation, to produce legitimate offspring for the service of the gods and the state. In the early Republic the *mater-familias* had dignity and some freedom but she was subservient to her father before marriage and her husband afterwards. When women, however, claimed the privilege of independence by 'three nights' absence', this led to the increase of divorce and disintegration of the family. Then the earlier puritanism of Rome was undermined by the introduction of Greek practices, and especially the corruption of youths. Although ideals of marriage and family life continued to prevail in many Roman provinces, the cities and seaports of the Roman empire became hotbeds of all kinds of vice. The lives of most of the Roman emperors themselves were notorious for licentiousness and cruelty, so that ascetic philosophies and new oriental religions came to serious-minded people as breaths of fresh air into a jaded world. Edward Gibbon, no friend of early Christianity, agreed that one of the principal reasons for the victory of Christianity was 'the pure and austere morals of the Christians'.

World-denying asceticism, with a depreciation of marriage, affected Christianity from the prevalent philosophies of the Hellenistic world, and there may have been other influences. How far ascetic Buddhism and Jainism had touched the western world is uncertain. Clement of Alexandria, at the end of the second century, wrote of 'those of the Indians who obey the precepts of Boutta'. He referred also to the celibacy of the 'naked philosophers', the Gymnosophists, perhaps Jain monks, who 'know not marriage nor the begetting of children'. Jerome, in the fourth century, supported his claim that virginity was a higher state than marriage by the dubious expedient of stating that virginity was so well esteemed among the heathen that some of them believed in virginal births. He said that there was a belief 'among the Gymnosophists of India that the Buddha, founder of their doctrine, was born of a virgin and emerged from her side.' But the Buddha's mother was married, and the Jains do not have such a belief.

There were many forms of Gnosticism which affected the early Church, developing trends from pagan religions and philosophies that regarded matter as inherently evil. Already in the New Testament we read of 'subversive doctrines' which 'forbid marriage and teach abstinence from certain foods' (1 Tim. 4, 3). Soon the Gnostics were powerful in the early churches, teaching a special 'knowledge' (*gnosis*) which was revealed to the 'spiritual' while others were merely 'fleshly' or 'material'. In the second century Marcion, who significantly came from the East where dualistic doctrines flourished, rejected the Old Testament with its God of Law, in favour of the God of Love revealed by Jesus and expounded by Paul with his contrast of flesh and spirit. But this love was spiritual only, and wedlock and procreation were condemned as the works of Satan.

By the end of the third century the extreme dualism of Manicheeism arrived in the West and profoundly influenced Augustine of Hippo. The founder of Manicheeism was Mani, who lived in Persia in the third century A.D. According to Augustine, Mani taught that there were two primary elements in the universe, God and matter. All goodness came from God, and all evil from matter, which was also called the Devil. But matter was not a god, and this belief differed from Zoroastrian ideas according to which the God of light and the God of darkness were twin spirits, and matter had been created by the good God. Contrary to the Zoroastrians,

Manichees held that matter was concupiscence, a 'disorderly motion in everything that exists'. Matter or concupiscence was female, the 'mother of all the demons', and the soul was imprisoned in it. The object of religion was to release the soul, and severe asceticism was practised to this end. The Manichean religion spread rapidly, was known in Rome by the fourth century and was influential in north Africa. How far its ideas affected later sects, such as the Cathari, the 'pure', of the twelfth century is uncertain, but the charge of Manichean has been levelled at many who have stressed the contrast of soul and body and have regarded sex as unclean.

THE TEACHING AND PRACTICE OF JESUS

Jesus was a Jew and his words on sex and marriage arose in the Hebrew background. Hellenistic and dualistic ideas began to emerge with Paul, yet Jesus lived in Galilee which was probably less strictly ritualistic than Judea, and the Gospels which relate his life and teaching were written in Greek.

Jesus affirmed the monogamic state as originally ordained by God, rather than a new teaching: 'From the beginning of creation he created them male and female', and Genesis showed that 'therefore a man shall leave his father and his mother, and shall cleave to his wife, and they shall be one flesh', so that they are no more two but one flesh. (Mark 10.6-9)

This statement was made in the context of divorce, and Jesus went on to say that 'what God has joined together, let not man put asunder.' Such criticism of divorce was against Jewish custom and the Mosaic law, which was said to have permitted divorce 'because of the hardness of your hearts'. Jesus said that 'whoever puts away his wife and marries another commits adultery against her', and the woman likewise 'if she puts away her husband and marries another, commits adultery.' These verses occur in Mark and Luke, and twice in Matthew.

It seems a high but difficult morality, and much has been made of these verses in the later Church's rigorous attitudes towards divorce. But in this instance Jesus seems to have been looking to the purpose of creation, and he took the divine pattern of the creation of man and woman in singleness and unity. In such a context, for a married man or woman to take another partner would be against the unity of their creation. Paul later declared that even he that is 'joined to a harlot is one body', nevertheless a Christian believer could be separated from an unbelieving partner. (1 Cor. 6)

The ideal set by Jesus for marital unity and indissolubility was paralleled by his other ideals. He criticized not only the act of lust but the lustful imagination: 'Everyone that looks on a woman to lust after her, has committed adultery already with her in his heart.' (Matt. 5.28f.) But the same concern with intention as well as act came in the accompanying statements about taking oaths and speaking truth, killing and anger, retaliation and turning the other cheek, love to friends and enemies.

Matthew's Gospel allowed one exception for divorce, 'except for the cause of fornication', though most scholars consider this a later addition and it did not appear in the other Gospels. Matthew also presented the disciples as exclaiming that if divorce was fornication it would be better not to marry. To this Jesus replied that some were born eunuchs, some were made so, and some became eunuchs for the kingdom of heaven, which would justify some kinds of celibacy.

The rigour of later church prohibition of divorce, in particular application of the ideal given by Jesus, was not applied to his other ideals. The Church has often sanctioned taking life in war or capital punishment, and it has insisted upon swearing oaths in civil and ecclesiastical courts of law. Yet while disregarding the literal application of the other ideal teachings of Jesus, the Church demanded literal obedience to the ideal against divorce. It has almost been regarded as the one unforgivable sin, though that mysterious fault was indicated by Jesus as sin against the Holy Spirit.

Most of the teachings of Jesus on morality, even the one about looking on a woman with lust, can be paralleled in the teachings of the rabbis who generally regarded the ideal marriage as one of permanent monogamy. The novelty of

Jesus seems to have been the freshness of his approach, sweeping aside quibbles about the law, and his open relationship with all manner of people. His moral attitudes are to be seen as much in his actions as in his recorded teachings. He was called 'a gluttonous man and a wine-bibber', notorious for being 'a friend of publicans and sinners'. The publicans were tax-gatherers for the public revenue, who were despised because of their rapacity and low morals, and also because their work involved contact with foreign rulers, which in modern times would have earned them the nickname of 'quislings'. The 'sinners' included both those who neglected the rigid rules of the Law and people of immoral life. All these outcasts needed God, and Jesus said that he had come 'not to call the righteous but sinners to repentance.'

When Jesus was eating in a Pharisee's house, a woman wept on his feet, wiped them with her hair, and anointed them with precious oil. She was known as a 'sinner', though it is not clear that her sins were sexual, but the Pharisee thought that Jesus should have known of her uncleanness. But Jesus spoke of her love, which would be pointed if she was a prostitute, and that she had much to be forgiven. 'Her sins, which are many, are forgiven; for she loved much.' (Luke 6.47)

In another instance a woman who had been 'taken in adultery' was brought to Jesus, with the inquiry whether she should be stoned, as the law of Moses prescribed. It may be recalled that tradition reported a similar query made to Muhammad (see p. 164). The law in fact decreed that 'both of them shall die', and there was no appearance of the offending male, even though the woman had been taken 'in the very act'. It was such hypocrisy, and the application of his principle against 'looking at a woman with lust', that gave a sting to Jesus's command, 'He that is without sin among you, let him cast the first stone.' When they had all gone, Jesus told the woman that he did not condemn her himself, but she should go and sin no more. (John 8)

It was the practice of Jesus, his association with outcasts, his concern for the despised and the morally sick, and his healing of the physically diseased, that impressed his contemporaries. Of his own personal life little is known. Was he married? It would be unusual for a Jew to be celibate, yet the thought of his marriage might be repulsive to some Christians who maintain the divinity of Jesus, which traditionally should include his full incarnation, in the 'flesh' which is the lot of all mankind. But the texts are silent, and there is no more evidence for his marriage than for other sexual involvement. Jesus attracted male and female disciples, and on the day of Pentecost the Holy Spirit seems to have descended on the twelve disciples and their female companions, 'the women, and Mary the mother of Jesus' (Acts 1.14). It was a spiritual ordination.

There were some ascetic teachings in the Gospels, and these parallel similar ideas among Jewish parties in the time of Jesus. It was a period of general social and political disturbance, when many people expected great conflicts or the end of an era, and some retired from the world to prepare for it. The best known of these were the Essenes, some of whose doctrines have become available since the discovery of the Dead Sea Scrolls from 1947 onwards. Whether John the Baptist or Jesus met such communities—perhaps the latter did in his forty days in the desert—is uncertain but the scrolls give pictures of some of the religious and cultural environment of early Christianity. Some of the Essenes lived in desert places, like Qumran, and others lived in towns. The historian Josephus said that some of them discountenanced marriage, but others did not and the fact that skeletons of women have been disinterred at Qumran shows that it was one of the less extreme communities. The Temple Scroll forbade sexual relations anywhere in the city of Jerusalem, and it was opposed to divorce. The Zadokite Document referred to members of the Qumran community who married and begat children, and told them to follow the precepts of the Law and its regulations for the relationship of husband to wife and father to child.

The Essenes were strict observers of the Law, and while some verses of the Gospel suggest a similar attitude, other verses are much less rigid, as in the treatment of the Sabbath. But Jesus was presented in the Gospels as demanding loyalty to himself, even before family ties. According to Matthew one should not 'love' father or mother 'more than me', and in Luke a disciple should 'hate' his father and mother, and wife and children (Matt. 10.37; Luke 14.26). The priority was loyalty to Christ, but that family life could follow it was shown by the fact that long afterwards Peter and other apostles took their wives with them on preaching tours (1 Cor. 9.5). The imminent expectation of the kingdom of God coloured all early Christian thought, but as time went on modifications of the belief became necessary.

PAUL AND OTHERS

Paul was a Jew, and proud of being 'a Hebrew of the Hebrews, circumcised the eighth day, as touching the law a Pharisee.' (Phil. 3.5) Yet he was also a Roman citizen, wrote all his letters in Greek, came from Tarsus in Asia Minor, and introduced some new ideas into the Hebrew background. Paul has often been loosely accused of inventing harsh teachings on sex and marriage, yet in some ways he was one of the most original and enlightened of early Christian teachers. It was Paul who said that 'to the pure all things are pure.'

The prophet Hosea, we have seen, spoke of Yahweh as the forgiving husband of Israel, and Paul took this idea further to show the divine sacrifice as an example to human marriage: 'Husbands, love your wives, even as Christ loved the church, and gave himself for it.' (Eph. 5.25) The ancient pagan symbolism of the sacred marriage, the union of the cult deity with his people, which had been purged by the prophets, was taken by Paul as a symbol of divine redemption and a pattern for earthly union.

Paul developed the teaching of Jesus, taken from Genesis, of the union of man and wife: 'Men ought to love their wives as their own bodies . . . For this cause a man shall leave his father and mother, and shall be joined to his wife, and the two shall be one flesh. This is a great mystery.' There was still, however, the traditional Hebrew primacy of the man and the subordination of the woman. Christians were to 'submit yourselves to one another', but wives should 'submit yourselves to your own husbands, as to the Lord. For the husband is the head of the wife, as Christ is the head of the church.' And again, 'Let every one of you in particular so love his wife even as himself; and the wife see that she reverence her husband.' Yet elsewhere Paul saw that in the new faith the distinctions of sex should be transcended: 'There is neither Jew nor Gentile, there is neither bond nor free, there is neither male nor female; for you are all one in Christ Jesus.' (Eph. 5.21f., Gal. 3.28)

Paul was an occasional rather than a systematic teacher, and his doctrines developed over some thirty years, according to the needs of his various audiences. His treatment of sexual intercourse in the first letter to the Corinthians was both realistic and central to his thought. Coitus was not a casual act, a mere exercise of genital organs but an expression of the whole personality. Physical intercourse unites, whether in marriage or in fornication, with due results. 'Do you not know that your bodies are members of Christ? Shall I then take the members of Christ and make them members of a harlot?' The two became one flesh in any copulation, so that 'he who commits fornication sins against his own body', which is the temple of the Holy Spirit (1 Cor. 6).

The following chapter contains some of Paul's chief teachings on sex, which were complex and have often been misunderstood. Two things must be borne in mind; the first is that Paul was answering questions sent by the Corinthian church, in a city that was notorious for immorality, and the second point was that Paul expected the coming of the end of the age and therefore his advice was for that period.

For the present, then, he preferred celibacy: 'I would that everyone should be as I am myself', because of 'the present distress', since 'the time is short.' Yet marriage was necessary for most people and 'to avoid fornication let every man have his own wife, and let every woman have her own husband.' It was good for the unmarried to remain single, but since passions were strong, if they could not control themselves they should marry, for 'it is better to marry than to burn.'

These counsels were given as Paul's own opinions, not permanent regulations, for they were 'concessions, not commands'. There was, however, the command of God against separation: 'To the married I command, yet not I but the Lord: Let not the wife depart from her husband . . . and let not the husband put away his wife.' The problem of a Christian partner with an unbelieving spouse was acute in the infant Church, and Paul first advised Christians to remain faithful to their partners, but also allowed the unbeliever to depart if necessary.

Paul admitted that he had no divine command about virgins, and gave his own opinion. He thought that an unmarried woman would care more for the things of the Lord, whereas a married woman 'cares for the things of the world, how she may please her husband', as a married man also seeks to please his wife. But this was not to exalt virginity permanently above marriage, and a virgin daughter or ward might be married: 'There is nothing wrong in it, let them marry.'

Paul himself seems to have had a tendency towards asceticism, though he said that he could have had the right to take a wife with him on his travels, as did Peter and the brothers of Jesus and other apostles. He preferred the single state, in the circumstances of the time, but he recognized that God called some people to celibacy and others to marriage.

He gave important approval to sexual relations within marriage. Married couples should not deny coitus to each other, since they had given each other power over their own bodies. 'The husband must give the wife her due, and the wife equally must give the husband his due. The wife cannot claim her body as her own; it is her husband's. Equally the husband cannot claim his body as his own; it is his wife's. Do not deny yourselves to each other.' (1 Cor. 7) Such teachings were fully Hebrew, and against the rigorism that later developed in the Church.

Paul denounced male and female homosexuality: 'women changed the natural use into that which is against nature; and likewise also the men, leaving the natural use of the woman, burned in their lust one towards another', and the wrath of God was revealed against all such unrighteousness (Rom. 1.26f.). And again, the law was against 'those that defile themselves with mankind.' (1 Tim. 1.10) This was in the Hebrew tradition.

213

Paul, however, can hardly be exempted from teaching the opposition of flesh and spirit, in a Greek philosophical rather than a Hebrew manner. 'The flesh lusts against the spirit, and the spirit against the flesh; and these are contrary the one to the other'; and again, 'those that are after the flesh mind the things of the flesh, but those that are after the spirit the things of the spirit. For to be carnally-minded is death; but to be spiritually-minded is life and peace.' (Gal. 5.17; Rom. 8.5f.) It has been argued that the 'flesh', in the Hebrew meaning was the whole personality seeking to live independently of God, and the 'works of the flesh', listed by Paul, included sexual lusts such as fornication, impurity, and licentiousness, but also referred to quarrels, envy, anger, selfish ambitions, and party intrigues. Further, the powers of darkness were spiritual, according to Paul, and his dualism was perhaps more like the Zoroastrian opposition of good against evil than the late Greek and Manichean warfare of spirit against matter. Nevertheless the introduction of such dualism was applied later to the opposition of flesh and spirit, body and soul, and it was used to justify asceticism and to depreciate sex.

The rest of the New Testament was similarly diverse and complex. The book of Revelation went to extremes in speaking of the elect who 'were not defiled with women', while on the contrary the anonymous author of the epistle to the Hebrews exhorted, 'Let marriage be had in honour among all, and let the marriage bed be undefiled.' (Rev. 14.4; Heb. 13.4)

The subjection of women to their husbands was taught in various places: they were the weaker vessels, told to learn in silence, not to teach in church, to be modest in ornament, and reminded that Eve was the first to be deceived into transgression though women could be 'saved through childbearing' (1 Pet. 3.1; 1 Tim. 2.9f.). On the other hand, women were called 'joint heirs of the grace of life'. They dispensed charity and hospitality, Priscilla was named as giving instruction in the faith, and she was mentioned before her husband Aquila. Women were among the first converts, as in most religions, and they probably formed the bulk of the faithful, many women being named in the Acts and epistles.

CHRISTIAN DIVERSITY

MONOGAMY AND LOVE

It has nearly always been considered that the Christian scriptures teach monogamy, and it may well be assumed if it is not clearly stated. Paul, writing to Timothy, said that the bishop or overseer should be 'the husband of one wife', and deacons also 'husbands of one wife'. To Titus he wrote that elders or presbyters should be 'blameless, the husband of one wife, having children that believe'. Some commentators have concluded that such church leaders should not contract second marriages after the death of the first wife, and such a rule was sometimes applied in the later Church, but the texts do not state this clearly. No similar regulations were made explicit in the Bible for ordinary Christians, though their monogamy may have been assumed. Some African Christian Independents claim that church members may be polygamists, since the New Testament does not clearly forbid them, but the growing ascetic tendency in the early Church would be against such an interpretation.

The unity of man and wife in 'one flesh', commended by Jesus and referred back to the original action of God at creation, seems to require a single lifelong union. Paul repeated this doctrine, and monogamous marriage was probably the general practice of Jews in the first century, though there were plenty of polygamous heroes in the Old Testament.

The New Testament says a great deal about 'love', of God to man, man to God, and husband to wife. There were several Greek words which could be translated love, such as *eros* for passionate emotion, or *philia* for friendship and dutiful affection, but the New Testament does not mention *eros* and for love it writes *agape*, which has been claimed almost as its own invention. Later theologians distinguished between *agape* and *eros*, as between sacred and profane, or 'seraphic' and 'carnal' love, and some modern writers have even claimed that man cannot have *agape* to God, since that would add something to God, but this seems legalistic and in the New Testament love is both a divine and a human activity.

Although *agape* seems hardly to have been used before the New Testament, it is interesting that it appeared in the

Greek Septuagint translation of the Old Testament, representing the commonest Hebrew word for love with a personal object. Even the passionate love of the Song of Songs, which might have been thought to be best represented by *eros* was translated by *agape* and its derivatives. *Agape*, then, could be used of sexual love as well as divine love, but it is used of many kinds of affectionate relationships.

It was with *agape* that 'Jesus loved Martha, and her sister, and Lazarus'; he 'loved' the rich young ruler, and at the Last Supper the disciple 'whom Jesus loved' reclined on his bosom. These were warm friendships, but there is no evidence, despite a modern claim, that any sexual love was involved. In the same manner God so 'loved' the world, and indeed God is 'love'. Man was told in the Law to 'love' God and his neighbour, and Jesus instructed his followers to 'love' their enemies, though in the epistles the emphasis was more on 'love of the brethren'. *Agape* was the central word of Paul's great hymn of 'love' in 1 Corinthians 13, formerly rendered in English by 'charity' but clearly meaning much more than is now implied by such a word.

Love was a central concept in the New Testament but later theologians did not pursue its implications in discussing sexual relationships, for these came to be regarded as merely fleshly and inferior to the life of the spirit as best exemplified by celibates. In the Middle Ages there did emerge the concept of 'romantic love' which eventually widened Western attitudes, but for a long time it was confined chiefly to literature. The troubadours of France composed lyric poems of exalted love to idealized women. In a different setting Dante celebrated his early love for the young Beatrice, seen as naked except for a crimson mantle in an early sonnet, and after her death the love became a semi-religious longing for the celestial figure of Beatrice in the *Divine Comedy*. But from such ethereal outpourings physical sex was almost absent. The glimpses that emerge of general patterns of married life in the Middle Ages and Reformation periods reveal little if any 'romantic' relations of husband and wife, though probably many couples were or became affectionately attached to each other. The dominant emphasis upon love in marriage, which could have been deduced from the New Testament, was long neglected by theologians, and its emergence in modern times owes much to romantic fiction, the work of laymen, and to social changes, especially the emancipation of women.

If *agape* should be the pattern for all behaviour then, following Biblical teaching, the union of husband and wife is its highest human expression. It follows that love, in the full Biblical sense, should determine marital relations, and not the legalism which for so long dominated sexual ethics. The belief that marriages are made in heaven would be consistent with the teaching of their being instituted by God, provided that they remained inspired by love.

Love demands an I-Thou relation between persons, whereas our relation to things is I-It. 'The primary word *I-Thou* can be spoken only with the whole being . . . So long as love is "blind", that is, so long as it does not see a *whole* being, it is not truly under the sway of the primary word of relation', so said Martin Buber. The wholeness of this relation finds its meaning in close encounter, for 'all real living is meeting.'[3]

Agape is supremely the love of God to man, 'we love, because he first loved us.' The relation implies a distinction as well as a union. 'The Father and the Son, like in being—we may even say God and Man, like in being—are the indissolubly real pair, the two bearers of the primal relation, which from God to man is termed mission and command, from man to God looking and hearing, and between both is termed knowledge and love.'[4]

It is not surprising that mystics used the closest symbol of relation, that of marriage, of their union with God, a union which also included difference so that dialogue could take place. At the end of the Bible was the vision of the divine wedding of the Lamb and the Church, and the latter was often later called the 'spouse of Christ' though the individual soul was also regarded as female and married to the male Lord, as nuns became the brides of Christ on taking their vows.

[3] M. Buber, *I and Thou*, Eng. trs. 1937, p. 16.
[4] Ibid., p. 85.

Physical sexual imagery was used of mystical relations, and men and women spoke of raptures, or being rapt, which come from the same root as raped. But the difference was in the willing consent of abandonment to God. One of the most famous instances of sexual symbolism in mysticism was the vision that Teresa of Avila had of an angel:

> In his hand I saw a great golden spear, and at the iron tip there appeared to be a point of fire. This he plunged into my heart several times so that it penetrated to my entrails. When he pulled it out, I felt that he took them with it, and left me utterly consumed by the great love of God.[5]

Since the time of Freud we can hardly avoid seeing the phallic symbolism of the great spear, though Teresa would have been horrified at such a sexual interpretation of her vision. She was ardent but celibate, seeking ever more severe renunciation of the world, and her longings were sublimated into divine love.

The use of such sexual imagery may be less shocking to us than to Victorians, and R. C. Zaehner commented:

> There is no point at all in blinking the fact that the raptures of the theistic mystic are closely akin to the transports of sexual union, the soul playing the part of the female and God appearing as male . . . The close parallel between the sexual act and the mystical union with God may seem blasphemous today. Yet the blasphemy is not in the comparison, but in the degrading of the one act of which man is capable that makes him like God both in the intensity of his union with his partner and in the fact that by this union he is a co-creator with God.[6]

THE ASCETIC EARLY AND MEDIEVAL CHURCH

As the Church moved out into the Gentile world, Jewish influence declined and Hellenistic and other ideas dominated. Paul's opposition of 'flesh' and 'spirit' was interpreted dualistically as sex against sanctity, even without

---

[5] *The Life of Saint Teresa*, tr. J. M. Cohen, 1957, p. 210.
[6] *Mysticism Sacred and Profane*, 1957, pp. 151 f.

Marcion. Paul's praise of virginity above marriage, even if only for a time, was taken as perpetual. The persecutions suffered by Christians for centuries encouraged a scorn of material comforts, and when worldly success did arrive its corruptions led to ascetic movements which despised sexual activity.

Sexual and marriage questions were not greatly debated in the early Church, but there was a world-renouncing attitude and Gnostic anti-sex feelings were shared by many of their opponents. Marriage was not condemned, for it was blessed by Christ and instituted by God, but it was regarded as inferior to virginity. Celibacy was the ideal for those who would be perfect, and clerical remarriage after the death of a spouse was forbidden. Tertullian, although a rigorist, expressed the general orthodox position when, attacking Marcion, he said, 'We do not reject marriage, but simply refrain from it . . . earnestly vindicating marriage, whenever hostile attacks are made against it as a polluted thing, to the disparagement of the Creator.'[7] Other writers agreed that wedlock should not be despised, though celibacy gave more time for devotion to God.

Early Christian teachers denounced pagan practices such as the exposure of unwanted babies, and they cared for orphans. But they regarded parenthood as full of troubles and recommended severe discipline to fathers. Some, like the mild Clement of Alexandria, while allowing that virginity was higher, recommended marriage for most people, and he wrote sympathetically of the happiness of a Christian home and disagreed with Paul that a man cannot please both his wife and God. His disciple, Origen, however, took literally the Gospel words about those who made themselves eunuchs and castrated himself, to his later regret.

There were some liberal thinkers, known now from the works of their opponents, who expressed different views. Mary, the mother of Jesus, was given as an example of the value of marriage, against the growing opinion that she remained ever virgin. The common view that virginity was

[7] For patristic and later references see D. S. Bailey, *The Man–Woman Relation in Christian Thought*, 1959.

preferable to marriage was disputed, and married clergy were said to be best, since the single life was an evasion of responsibility. Three of such writers were viciously attacked by Jerome, one of the most learned and scurrilous patristic theologians, who advocated extreme asceticism.

The clergy, supposedly specialists in sanctity, soon came under pressure to remain celibate, especially in the western church, despite the married example of Peter and Pauline instructions to married deacons and bishops. The eastern Orthodox Christians were more moderate in allowing priests to marry, though they could not take a wife after ordination and if a priest became a bishop his wife was obliged to retreat to a nunnery. The permission of priestly marriage remains in the Orthodox churches, though bishops must be celibate and therefore they are generally promoted from the ranks of monks.

In the western church from the fourth century attempts were made to impose sexual abstinence on bishops and priests, and others who served at the altar, though in earlier years excommunication had been proposed for anyone who objected to a married priest celebrating the Mass. Successive councils, however, in trying to persuade married clergy to have their wives 'as if they had them not', revealed that the regulations had met with considerable opposition. Efforts against the marriage or concubinage of the clergy continued right up to the tenth century, and even later in some places, and clerical marriage was re-established as the general rule at the Reformation. Today the Roman Catholic church is the only religion in the world to insist upon universal and compulsory celibacy for all its clergy.

A curious compromise in sexual asceticism was the 'spiritual marriage' of a couple who shared the same house, and often the same bed, ostensibly in continence and as brother and sister. Paul had spoken, somewhat mysteriously, of a man's behaviour to 'his virgin', but he recommended marriage if there was anything uncomely (1 Cor. 7.36). The practice of co-habitation seems to have been derived from paganism, and in the patristic age hermits were often accompanied by a female hermit, and celibate clergy had a 'bride of the soul', who might become the housekeeper or even the mistress. Jerome asked, 'Whence come these un-wedded wives, these new types of concubines, nay, I will go further, these one-man harlots? . . . They call us suspicious if we think that anything is wrong.' Bishops and councils forbade the 'so-called sisterly co-habitation', but it persisted for centuries, and when celibacy was finally enforced in the West the spiritual partnership tended to be replaced by clerical concubinage.

The general asceticism of the Church's teaching was taken to extremes in the development of monasticism, as a reaction against worldly corruption, whether in solitary eremitism or collective coenobitism. Antony of Egypt was one of the earliest hermits, who towards the end of the third century retired completely into the desert. But despite his self-torture the hermit could not leave his passions behind him, and the temptations of Antony became famous, in which he was said to have fought with demons under the guise of wild beasts, good Freudian symbolism. Jerome described his own experiences in the desert in the customary vivid terms in which he attacked sex. 'Though in my fear of hell I had condemned myself to this prison-house where my only companions were scorpions and wild beasts, I have often found myself surrounded by bands of dancing girls . . . The fires of lust kept bubbling up before me when my flesh was as good as dead.'

Both monasticism and clerical celibacy brought problems of their own making. It was understandable that monks and nuns should be unmarried, for living in community would have made it difficult to be otherwise, and this was the practice in Buddhism and other monastic systems. Nevertheless the encouragement of adolescent vocations on a wide scale brought many tensions to growing young men and women, and vows could not be ended as easily as they could in Buddhism. No doubt accusations of monastic and clerical immorality and homosexuality, made by Boccaccio and the Reformers, were exaggerated but there was enough truth in them to strike home.

Moreover their pastoral work brought the priests fully into touch with the sexual struggles of their parishioners, and this was intensified by the development of the confessional. Men and women, boys and girls, had to confess their sensual thoughts and temptations to single men, and with the Roman genius for organization detailed and prurient inquiries were worked out. Between married couples it might be asked how many times a night they had sexual intercourse, or whether coitus was performed in other than the natural manner. Penances were given for risqué jokes, reading obscene books, or touching parts of a woman's body, going on to suggestions of variations in sexual behaviour in sodomy and bestiality. Not only the penitent but the priest must have been affected by constant inquiry into forbidden sexual behaviour, and beneath much of it lay the assumption that sex belonged to the lower nature, if it was not positively evil.

Again and again theologians spoke of the dangers and inferior nature of sex, but Augustine, Bishop of Hippo in north Africa in the early fifth century, was one of the most influential and disastrous teachers on the subject. His personal history affected his teaching, for he had a mistress for fifteen years, yet became a Manichean believing in the evil of the flesh. On return to Christianity he prayed, 'give me chastity, but not yet', sent his mistress away but took another, whom he also abandoned in favour of celibacy and conversion, 'so that I neither sought a wife nor any other of this world's hopes.' After such experiences Augustine's views of sexual relations were always dualistic. He was offended by the 'bestial movements' and 'lust' of sexual intercourse which, with childbirth, takes place between the organs of defecation and micturition, as Augustine put it in his coarse way.

Augustine believed strongly in Original Sin and Predestination. The sin came from the Fall of Adam, as a result of which man suffered from a hereditary moral disease, and was also subject to the inherited legal liability for Adam's sin. The whole human race was one mass of sin, from which God had predestined some souls to receive his unmerited mercy, but others went to hell, including unbaptized babies. Original Sin was virtually equated with sexual emotion, so that every act of coitus was intrinsically evil, and therefore every child was conceived by the 'sin' of its parents. Marriage could not take away the evil of sexual desire, thought Augustine, but it could divert it usefully to procreation, so that the only justifiable sexual act was one which aimed at producing children, a notion that infected Christian teaching for many centuries. Augustine's idea that the pleasure or 'lust' of sexual intercourse was sinful, the concupiscence of flesh against spirit, 'has had a most disastrous influence upon much of traditional Christian ethics.'[8]

LAY MARRIAGE AND PROBLEMS

What were ordinary Christian men and women doing in Europe, while their male celibate teachers pontificated on coitus as defilement and a hindrance to the highest service of God? They were flirting and loving, marrying and giving in marriage, as people had done since the world began. Probably most most laymen knew little of the arguments by which theologians defined the sex act as evil, if not always morally evil, but through priests and friars they must often have been aware that in the official Church's view physical intercourse was regarded as unworthy for a religious person, or secondary to the true will of God, supposedly linking married couples to the beasts rather than to the saints.

Marriage was a central concern of the laity, and although male priestly theologians might aim at celibacy for themselves, they could not resist defining the behaviour of their followers. According to Roman law the consent of the two parties was sufficient for a legal marriage, and while other European peoples had differing customs the Church eventually took over and imposed the Roman legislation. A few teachers thought that a priest and a church were necessary for a valid Christian ceremony, but it was soon established that these were only accessories and not essential to a true marriage. The ministers of the union, in Christian terms, were the bride and bridegroom themselves, and their consent sealed the contract. Although the Council of Trent in the sixteenth century made priest and church essential, and

---

[8] J. Burnaby in *A Dictionary of Christian Ethics*, ed., J. Macquarrie, 1967, p. 23.

was followed unwittingly by some Protestants, yet the consent of husband and wife remained the basis of both secular and Christian marriage.

By the end of the twelfth century there arose the definition of special symbols called sacraments, and after some uncertainty their number was fixed in the western church at seven. A sacrament was 'an outward and visible sign of an inward and spiritual grace', and there could be little doubt that marriage fulfilled this definition, though there was debate as to the proper 'sacramental moment' at which the grace was bestowed. At the Reformation most teachers rejected marriage as a sacrament, because it had not been instituted by Christ himself, as had the two Dominical sacraments, baptism and the Lord's Supper. Yet the claim for marriage to be recognized as a sacrament remained strong, was widely accepted, and was a help to the stability of marriage.

The Christian view of marriage, however, did not depend solely upon consent, and the Hebrew element in the faith came to the rescue once more. Jesus, quoting Genesis, had said that husband and wife become 'one flesh', and Paul had said that the partners should render each other their due. Despite some reluctance, especially from Augustine, the early Christian fathers accepted that coitus was essential to marriage and not a mere accessory. The consummated union alone could represent the human counterpart of the union of Christ and his Church. This view was strengthened by regarding the primary aim of marriage to be the procreation of children, following the biblical injunction to 'be fruitful and multiply'.

In many other ways the life of the laity as well as that of the clergy was regulated by sexual restrictions. From the second century, to the biblical prohibitions of adultery, fornication, and homosexuality, were added abortion and infanticide. In the eastern church Basil of Cesarea in the fourth century gave disciplinary rulings on sexual offences which included fornication, abortion, homosexuality, bestiality, rape, bigamy, second marriages, desertion, and incestuous or other forbidden unions. In the west, Theodore of Canterbury in the seventh century issued a Penitential for treatment of offenders against morality, which included male and female homosexuality, fornication, adultery, incest, bestiality, and male and female masturbation. Certain marital practices were also condemned: fellatio, coitus from behind, anal coitus, and coitus during menstruation. In the Middle Ages a connection was made between homosexuality and heresy, and when Manichean ideas from Bulgaria infiltrated into France allegations of sodomy were made against some of the Cathari and it was called Bulgarian or buggery.

Thomas Aquinas in the thirteenth century gave the fullest treatment of sexual diversions, in discussing the cardinal virtue of temperance and its contrary vice of lust. Everything had its proper end and sin was that which contravened the order of reason. The appointed end of sexual intercourse was procreation, and it was against reason when the right generation was rendered impossible as in contraception, incest, adultery, or rape. Kisses and caresses were innocent in themselves, but they became mortally sinful if their purpose was forbidden pleasure. Nocturnal 'pollution', as it was significantly called, was not sinful in itself but it might be caused by lustful thoughts or intemperate eating and drinking.

The most serious 'unnatural' lusts, for Aquinas, were masturbation, bestiality, homosexuality, and unusual modes of intercourse. Masturbation was the least evil but it was thought to be against reason and not helping procreation, and so more grievous than incest, adultery, rape, or simple fornication. The horror of male masturbation, which still appears in Roman Catholic manuals of sexual ethics as a 'grave moral disorder', was partly because of the 'misuse' or 'waste' of the precious semen, whereas female homosexuality and masturbation were ignored or dismissed as mere feminine lewdness.

Paradoxically, as some writers have noted, prostitution was uneasily recognized by theologians, as if some safety valve was needed by repressed clergy and laity. Augustine condemned this profession but continued, 'Yet remove prostitution from human affairs and you will pollute all things with lust', so that harlots were 'a lawful immorality'.

225

Aquinas made the same point: 'Take away the sewer and you will fill the palace with pollution . . . Take away prostitutes from the world and you will fill it with sodomy.' This seemed to allow prostitution as a corollary to celibacy, to establish a double standard of morality, and to express a grossly male view of sexuality. [9]

Against all the clerical attempts to regulate their lives the laity expressed reactions. Boccaccio in Italy, Rabelais in France, and Chaucer in England told irreverent stories of the immorality of clerics, monks, and nuns, and revelled in the sexual adventures of lay people. There was distortion alongside shrewd observation of sexual tensions, fantasy beside cold reality, and agnosticism as well as desire for religious reform. Eroticism rarely appeared in art because of early reactions against the overtly sexual features of much pagan classical sculpture and painting. But erotic images remained scattered over Europe, and medieval craftsmen often managed to use phallic imagery. Jewelry and furniture sometimes showed erotic scenes, and illustrated manuscripts depicted sexual acts. The Bayeux Tapestry had an explicit nude scene of invitation to copulation, but nude scenes of the Garden of Eden in devotional Hour Books were of a more spiritualized hedonism. At the Renaissance humanism brought renewed interest in the sensuality of the classical world and countless depictions of the human form and expressions of love.

VIRGIN BIRTH AND MOTHER

The Virgin Birth, or Virginal Conception, of Jesus came in the Gospels of Matthew and Luke, but was not mentioned in the oldest Gospel, Mark, the latest, John, or any of the Pauline or other epistles. It could hardly therefore have been an essential item of faith for the first Christians. But even in Matthew and Luke there was no suggestion that the normal processes of procreation and birth were unworthy of Jesus, the material world unclean, or virginity superior to the

[9] D. S. Bailey, *The Man–Woman Relation in Christian Thought*, pp. 161 f.

married state, as so many later teachers thought. In Luke's Gospel Mary was represented as conceiving through the power of the Holy Spirit coming upon her and over-shadowing her, rather like the birth of Samuel in the Old Testament. In Matthew's Gospel the conception was said to fulfil a prophecy of Isaiah, 'behold a virgin shall be with child', or more accurately 'a young woman'. In Isaiah the prophecy had been given to encourage king Ahaz by saying that a young woman was having a child but before it grew up his enemies would be destroyed. Both Matthew and Luke traced the genealogy of Jesus through Joseph as father, and Luke later referred to the 'parents' of Jesus. In Mark there was a reference to the brothers and sisters of Jesus, four brothers being named, and the ordinary words for brothers and sisters were used (Mark 6.3). It would have been natural, in the Hebrew setting, for Joseph and Mary to have had other children than Jesus, but in the later church ascetic tendencies militated against such a notion.

Apocryphal infancy stories delighted in multiplying the miraculous elements in the Nativity story and, with that prurience which often marks such narratives, the Protevangelium of James in the second century named a midwife, curiously called Salome, who sent into the 'cave' after Jesus had been born and 'made trial' to see if Mary was still a virgin. Mary's parents were here named for the first time as Joachim and Anna, and such highly embellished tales have been popular down to modern times and in the Middle Ages they provided many subjects for artists. Islam in the seventh century reflected some of the popularity of tales of Mary, and the Qur'an told of Mary being brought up in the temple and conceiving Jesus by the powerful Word of God. Mary was the only woman mentioned by her proper name in the Qur'an.

By the fifth century at latest in Christian circles the per-petual virginity of Mary from being a popular opinion had attained dogmatic status, but it raised the question whether her union with Joseph could be called a marriage. Joseph was represented as an old man, though there was no evidence for this in the Gospels and it might seem more likely that he was a young lover. Origen already in the third century had suggested that the brothers of Jesus were his half-brothers, children of Joseph by a previous marriage, again without evidence, and this became accepted as fact. Behind such notions was the thought that the conjugal bed was unclean, yet since marriage was a divine institution, and the union of Joseph and Mary had been commanded by an angelic messenger, their union had to be depicted as complete. Augustine therefore affirmed that Mary was a true wife, according to the theory of consent for valid marriage, and Aquinas later asserted that Mary and Joseph 'consented to the nuptial bond, but not expressly to the carnal bond, except on the condition that it was pleasing to God.'

In the fourth century a Christian sect, the Collyridians, was said to have offered cakes to the Virgin Mary after the pagan manner of worshipping the goddess Ceres with cakes. But the orthodox cult of the Virgin was slow in developing, since devotions were offered to martyrs but Mary had died peacefully. Yet the theologians had early quoted her as an example of holy life in the new dispensation, and Tertullian quoting the contrast which Paul had made between Adam and Christ placed alongside them Eve and Mary, to complete the new humanity.

A great impulse to the cult of Mary came with the victory of the Greek term Theotokos, God-bearer. It had been used of Mary since the time of Origen, but in the fifth century it was attacked by Nestorius, Patriarch of Constantinople. He said that such a term was incompatible with the full humanity of Christ, and that Mary should rather be called Christotokos, Christ-bearer. Nestorius was defeated in controversy by Cyril of Alexandria and he retired, though his missionaries spread across Asia as far as China. The orthodoxy of the title Theotokos became established, though in the West it was generally translated by the slightly different Latin term Dei Genetrix, Mother of God.

Devotions to Mary flourished and developed over the ages. Her own Immaculate Conception, the dogma that 'from the first moment of her conception the Blessed Virgin Mary was . . . kept free from all stain of original sin', was long debated. Bernard, Aquinas, and others denied it since Mary had no doubt been conceived in the natural way, but

228

the notion gained ground and it was defined as a dogma by Pius IX in 1854. As with the doctrine of the Virgin Birth of Jesus, the dogma of the Immaculate Conception of Mary does not, openly at least, consider sex as unclean, but it seeks to remove from Mary all trace of original sin. Similarly the dogma of the Assumption of the Blessed Virgin Mary, that having completed her earthly life she was in body and soul assumed into heavenly glory, parallel to the Ascension of her Son Christ, became decreed by the Roman Catholic Church in 1950. The doctrine was unknown in the early Church but appeared from the fourth century, was held in the East as in the West, and was championed by theologians like Aquinas.

The popularity of the Virgin Mary has been very great, in Orthodox and Catholic churches, that is the majority of Christendom. Her cult was rejected at the Reformation, though some Protestants have felt the need for a female element in the deity, and psychologists like Jung have emphasized the importance of feminine symbols of divinity, the 'archetypes' of the female *anima* to correspond to the male *animus*. Most religions seem to need a goddess or female side of deity, though with the strange exception of Islam. In early Jewish-Christian circles there was some interpretation of the Holy Spirit as female, the dove or the personification of Wisdom. This failed, but though the Holy Spirit in the Trinity remained a somewhat vague figure, the cult of the Virgin-Mother developed rapidly. If she could have been called God the Mother, that would have completed the divine family with God the Father and God the Son.

Medieval Christian devotion achieved the logically impossible feat of adoring Virginity and Motherhood in the same person. But surely she could have been revered as the supreme example of motherhood, without the contradictory notion of virginity? To some degree the elevation of Mary compensated for the inferior status of womanhood held by ascetical teachers of the Church, and her high place was the result of upsurges of passionate devotion.

REFORMATION CHANGE AND CONSERVATISM

Some sexual questions were not debated at the Reformation

in western Europe, and there was little attempt to work out a theology of sex. Although the Reformers appealed to Scripture, they interpreted it in conservative ways, and not until modern times has the critical approach to both Scripture and Church tradition enabled Christians to make fresh interpretations of sexual relations. Puritanism and Victorianism are both now regarded as movements of sexual repression and prudery, but their roots go right back through the ages as far as the early Church. The dualistic opposition of flesh and spirit was maintained at the Reformation, despite other changes, and the place of woman was still subordinate to that of man. Anti-material phrases still linger, when at Christmas we sing, 'Lo, he abhors not the virgin's womb', as if it were unclean; or at a more banal level, 'little Lord Jesus no crying he makes', which would have made him an unhealthy baby.

The Reformers did, however, immediately attack two sexual topics: clerical celibacy and monastic vows of continence. Marriage was exalted as a divine institution, for everybody, and celibacy was condemned as contrary to divine law. The Reformers criticized the clerical immorality and concubinage which had brought discredit on the Church, and had made it the laughing-stock of satirists like Boccaccio.

Zwingli declared that marriage must be lawful for everyone, since Scripture nowhere forbade it, and it was sinful for priests and monks not to marry if they knew they could not keep the vows of chastity. Luther advised those about to be ordained not to swear continence, and he said that a priest who had a concubine should marry her, despite the pope's displeasure or public opinion, since in God's sight they were already married. Vows of chastity, he said, originated in the delusion that asceticism can win the favour of God, but it is impossible to resist temptation otherwise than in the manner which God has ordained. Calvin was more cautious, and condemned only such celibacy as was undertaken by those who could not keep the vows. But he held that the Church had no right to condemn what Scripture showed to be an open choice.

Luther married the former nun Catherine von Bora eight

years after the publication of his famous ninety-five theses, but despite his outspoken comments on other themes he followed some of the negative ideas of previous theologians on marriage. He regarded it as 'a hospital for the sick', an antidote for the incontinence which troubles every man, and sexual intercourse had some shame in it which came from the Fall of man. Calvin regarded coitus as honourable and holy, and he attacked Jerome for branding as unclean what Scripture had given as a symbol of the union of Christ and the Church. But Calvin, characteristically, was uneasy about sexual pleasure, which he felt must have some evil in its immoderate desires that he also traced back to the Fall. Luther thought that celibacy should be shunned since it could be maintained only by 'peculiar' (*seltsam*) persons, perhaps one in a thousand. Under the outward pretence of celibacy or virginity there often burned evil desires. But Calvin accepted that virginity was a virtue, and it was even superior to marriage if it was not adopted under compulsion.

It is well known that Luther reluctantly sanctioned the bigamous marriage of Philip of Hesse, because he detested divorce and believed that the Church had the power to make exceptions to the rule of monogamy in extreme cases. After all, the popes had often annulled marriages, and the biblical patriarchs had been polygamous, and although their example was not to be followed by Christians in ordinary circumstances, there was no biblical command against it, an opinion with which some modern African Christians would agree. Calvin, however, regarded polygamy as contrary to the divine institution of monogamous marriage and he thought, unusually for a theologian of his time, that the Hebrew patriarchs had been wrong to marry more than one wife.

In England Thomas Cranmer played the major part in Henry VIII's divorces, advising him to consult the universities for an assurance on the invalidity of his first marriage instead of waiting on the Pope, annulling his marriage to Catherine in 1533 and three years later that to Anne Boleyn. Cranmer himself was first married while a fellow at Cambridge, and although he was not yet ordained the university rules of the time which required celibacy of dons lost him the appointment. However, his wife died a year later in child-birth and Cranmer was re-elected to a fellowship. He was ordained priest in 1523 and made Archbishop of Canterbury by Henry in 1532. Cranmer had just secretly married Margaret Osiander, the niece of a continental reformer, and there had been a move among English clergy to marry in the expectation that clerical celibacy would shortly be abolished by Henry's reforms, but this did not happen. The legal recognition of priestly marriage in England had to await injunctions of 1547 in the reign of Edward VI and 1559 under Elizabeth.

Although clerical marriage came to be approved by Protestants, the attitude of theologians to marriage and sex in general still reflected belief in the dualism of flesh and spirit, with honours to the latter. The Anglican Book of Common Prayer, used in various versions from Cranmer's time, continued to inflict dualistic sentiments upon many of the English-speaking peoples for nearly four centuries. The form of Solemnization of Matrimony stated from the 1549 version onwards that it had been 'instituted of God in the time of man's innocency', but continued gratuitously that it was not to be undertaken 'to satisfy men's carnal lusts and appetites, like brute beasts that have no understanding'. This was a libel on animals, many of which are less sexually active than mankind.

Three reasons were given for the institution of marriage; it was primarily for 'the procreation of children', secondly 'for a remedy against sin, and to avoid fornication; that such persons as have not the gift of continency might marry, and keep themselves undefiled'. Only thirdly was marriage declared to be 'for the mutual society, help, and comfort that the one ought to have of the other, both in prosperity and adversity.' Concluding exhortations in the ceremony, quoting Paul and Peter, told husbands to honour their wives as weaker vessels, and wives to submit themselves to their husbands, whom they had promised to obey. Only in the revised prayer books of 1927 and 1928, rejected by Parliament but used in many churches, did most of these phrases disappear.

Marriage was said to have been ordained for procreation

and as a remedy against sin, but if a child was born the Prayer Book service of Baptism declared in its opening sentences that 'all men are conceived and born in sin'. You could not win, and a prayer asked that 'all carnal affections may die in him', which if answered would have made the child a eunuch indeed. The 1928 revision was more kindly, saying that 'all men are from their birth prone to sin', and praying that all 'evil' desires might die in the child. Curiously the Methodists, the largest English Free Church, which had followed but revised the Anglican orders of service, continued to pray until 1975 for the baptized baby 'that all things belonging to the flesh may die in him,' which if answered would have destroyed the church membership. As R. C. Zaehner remarked, 'At baptism Christians solemnly renounce the world, the flesh, and the devil. Of course they do nothing of the kind, for to all who are not Buddhists or Manichees the world and the flesh have their due place in their lives.'[10]

The appreciation of marriage and sex slowly progressed in the Protestant world, though with some opposition or reservation. George Herbert in the seventeenth century thought that a country parson should remain celibate since his ministry required 'the best and highest things', and William Law in the eighteenth century mocked at 'reverend Doctors in sacerdotal robes making love to women.' However, in the seventeenth century John Donne, poet and later Dean of St Paul's, showed in his *Songs and Sonnets* and *Elegies*, a delight in love and sexual intercourse, although they were probably written before his ordination and his 'Holy Sonnets' came later. But it was the seventeenth-century theologian and bishop Jeremy Taylor who provided some of the most significant advances in sexual teachings.

In his books of practical divinity, with the forbidding titles *Holy Living* and *Holy Dying* and *Ductor Dubitantium*, Taylor showed a rare breadth of understanding of sexual relations. Virginity, he said was not 'purer' than marriage, and though it might be useful for some persons it was not better in the service of God. Paul's statement that the

[10] *Concordant Discord*, 1970, p. 342.

233

married cared more for worldly things was turned round to show that they were freed from many carnal temptations and had a more varied piety. In fact, 'single life makes men in one instance to be like angels, but marriage in very many things makes the chaste pair to be like Christ.' Marriage was the best of friendships, 'a union of all things excellent'. The husband was to be 'paternal and friendly, not magisterial and despotic', though, following tradition, the wife was to be submissive and engaged in domestic and nursery employment. Sexual intercourse, said Bishop Taylor, should be 'moderate, so as to consist with health', but 'without violent transporting desires, or too sensual applications'. Yet while its primary purpose was still thought to be procreation, it was also to be used 'to endear each other'. The silent wives of such theologians were probably responsible for their more positive and loving views of sex.

Jeremy Taylor was remarkable for justifying coitus during menstruation and pregnancy, for the former had been forbidden by Leviticus and the latter by theologians like Jerome who said that the very beasts abstain from copulation at such times. Taylor retorted that we are not mere animals, nor are Christians under the Mosaic law, and since Paul had said that husband and wife should not defraud each other from the use of the other's body, they might lawfully come together during menstruation and pregnancy when this was necessary.

The Reformers did not usually discuss other sexual acts, which had been condemned by celibate medieval moralists like Aquinas, but Taylor said that sins so-called 'against nature' were not worse than other sins such as adultery and fornication, which had been expressly forbidden by divine command. He was less permissive than some earlier casuists on the toleration of prostitution, and he attacked the notion that male nature needed a safety-valve while recognizing that prohibition of marriage had aggravated the evil. But the real cause of prostitution, said Taylor, was that 'men make necessities of their own, and then find ways to satisfy them'.

After such limited discussions of sexual problems in the sixteenth and seventeenth centuries, little else of importance appeared for nearly two hundred years, until the explosion of modern times. Writers who discussed marriage followed their predecessors, and on other sexual matters there remained the influence of the long ascetic tradition of the Church, which has been sketched above and in a few modern studies, with its dualism of flesh and spirit.

The Puritan and Victorian attitudes of repression of sex had a long ancestry, but they were also partly reactions to periods of what appeared to them to be licentiousness, or what today would be called permissiveness. Repression led to revolt, action led to reaction. This could be seen in Roman Catholicism where the licence of Renaissance Italy, notably under the Borgias, led to revulsion. The naked beauty shown in paintings and sculptures by artists patronized by the Papacy was too much for the Counter-Reformation. The Council of Trent condemned pictures that excited lustful feelings and Pope Paul IV had Michelangelo's nudes in the Sistine Chapel covered with draperies. In France the Jansenists, though critical of some views of the Counter-Reformation, were even more harsh in moral judgements.

There were similar swings from one extreme to another in Protestantism. The excesses of the Anabaptists in Germany encouraged moral strictness in Lutheranism and even more in Pietism. In Britain and America attitudes of English Puritan and Scots Kirk harshness, followed by Restoration licence, led on to Victorian prudery and modern revolt. There were many cross- and counter-currents, theological as well as moral. The dominance of both Evangelical piety and Broad Church laxity in the eighteenth century, was met by High Church revival in the nineteenth century praising clerical celibacy and exalting virginity. Then new forces gathered together to produce demands for revised sexual teachings today.

MODERN TIMES

Various factors contributed towards new assessments of sex in modern times, and some of them affect all religions and will be considered briefly in the closing chapter. But especially significant for Christian attitudes in the past hundred years has been the development of a critical

235

approach to both Scripture and tradition. Other religions will doubtless have to face such criticism of their sources in the future, but at present Christianity has been the most influenced, beginning with Protestantism and now openly seen in Roman Catholicism.

Many ordinary Protestants, clerical and lay, may now be more aware of the sexual ideas of psychologists and feminists than of those of the early and medieval church fathers. So when they look for religious guidance in the perplexing problems of sex, it is easy to jump over the ages and go back to the Bible, to discover Hebrew naturalistic attitudes to sex. From there criticism may be made not only of Old Testament polygamy but even of the dualism of Paul. Hence it is now often proclaimed and preached, incredibly, that Christianity is the most material of religions and, within limits, the most natural in sexual matters. Roman Catholics, even in liberal countries, may find it harder to take such a cavalier attitude to history, but the tendency may be observed at work. The more informed theologians, Protestant and Catholic, have performed uneasy contortions that indicate the tensions under which they are placed today.

Many pronouncements have been made by single or group church councils on sexual matters, and a few examples may be quoted. Various Anglican ten-yearly Lambeth Conferences since their inauguration in 1867 considered the nature of man, the requirement of purity, and the indissolubility of marriage. But not until 1958 were sexual relations fully discussed and, despite some tortuous arguments, the traditional sexual teaching of the Church was broken at several points. The absolute primacy of procreation was rejected, the personal value of sexual intercourse was emphasized, and any notion of its evil was condemned. Most notably, contraception by methods 'admissible to the Christian conscience' was approved as a means of family planning.

Other Anglican reports, especially one on *Marriage and the Family in Britain Today*, of 1974, spoke of the sexual act clearly and naturally, and sexuality was seen as directing much of human personality rather than being confined to coitus. But a report of an Archbishops' Commission,

Marriage and the Church's Task, issued in 1978, which would have allowed some recognition of divorce and remarriage in church, was rejected by the General Synod, perhaps because of a feeling that to change its marriage discipline would weaken Christian witness in a divorce-prone culture. Other churches, however, in Europe and America, had relaxed their rulings against divorce and remarriage, and they often remarried Anglicans and Roman Catholics in their services.

In Roman Catholicism numerous theologians tried to handle moral issues with more freedom than in the past but were restricted by rigid encyclicals and decrees. The encyclical *Casti Connubii* in 1930 reasserted the traditional position, but there were hints of a change in a document of the Second Vatican Council, *Gaudium et Spes*, issued in 1965. This stated that 'marriage is not insituted merely for procreation', and that 'the family is a kind of school of abundant humanity'. However, hopes of some recognition of artificial contraception, which was widely practised, were dashed by the publication of the encyclical *Humanae Vitae* by Pope Paul VI in 1968. It began well by approving 'the new understanding of the dignity of woman' and 'the value of love in marriage and of the meaning of intimate married life in the light of that love'. But the encyclical rejected the majority report of its preparatory commission: 'The conclusions arrived at by the Commission could not nevertheless be considered by Us as definitive and requiring absolute assent, neither could they dispense Us from the duty of examining personally this serious question.' Therefore the papal authority was imposed, for 'the Church is competent in her Magisterium to interpret the natural moral law', and the Church 'has always provided consistent teaching on the nature of marriage, on the correct use of conjugal rights'. From this it was decreed that husband and wife 'are not free to act as they choose in the service of transmitting life'. The Church forbade 'any action, which either before, at the moment of, or after sexual intercourse, is specifically intended to prevent procreation—whether as an end or as a means'. Equally condemned was sterilization, of the man or woman, and 'above all, direct abortion'.

The church allowed married people to take advantage of natural cycles and 'use their marriage at precisely those times that are infertile'. One reason given for rejection of contraception was that 'a man who grows accustomed to the use of contraceptive methods may forget the reverence due to a woman, disregarding her physical and emotional equilibrium'. That this might happen even more by having intercourse in infertile periods, was not mentioned.

Controversy exploded over this encyclical, and newspapers received thousands of letters of protest from Catholics. Some priests preached against it and were disciplined, and others disregarded it in practical advice. The Orthodox Ecumenical Patriarch supported the Pope, but the Archbishop of Canterbury remarked that such teaching was 'widely different from that of the Anglican Communion'. Similar Roman Catholic official rigidity appeared in a *Declaration on Certain Questions concerning Sexual Ethics* published in 1975. This condemned 'sexual union before marriage' even where there was 'a firm intention to marry and an affection which is in some way conjugal', because traditional moral doctrine stated that 'every genital act must be within the framework of marriage'. Homosexuality was denounced as lacking 'an essential and indispensable finality', for homosexual acts were 'a serious depravity and even presented as the sad consequence of rejecting God'. Masturbation was regarded as 'a grave moral disorder', and despite the fact that psychologists and sociologists claimed it to be a normal phenomenon of sexual development 'this opinion is contrary to the teaching and pastoral practice of the Catholic Church'. Masturbation, it was said, had been considered as a 'seriously disordered act' from a decree of 1054, 'even if it cannot be proved that Scripture condemns this sin by name'. All three practices were judged illegitimate because they lacked 'the full sense of mutual self-giving and human procreation in the context of true love'.

Differing views were widely held at a less official level, and an example of a reply to both *Humane Vitae* and the 1975 *Declaration* appeared from a Roman Catholic doctor, Jack Dominian. He began by defining sexual morality on the basis of 'the concept of person, in terms of human wholeness, and love'. A return to New Testament teaching on love would raise Christian views on sexual relations to the highest plane. Traditional sexual moral teachings needed to be re-examined in this light and authoritarian claims should be modified by claiming the rights of laymen and laywomen. Sexual problems were considered in turn: sexual pleasure, adolescent masturbation, pre-marital sex, sex in marriage, and contraception, with associated problems which should all be judged by the principles of love and personality. It was concluded that the Christian sexual ethic demanded 'a positive acceptance of human sexuality' in relationships of love.[11]

Christian attitudes to sexual questions in modern times were plainly divided, even within the Roman Catholic Church. There was general agreement on the ideal of monogamy, though with different opinions on divorce and remarriage. Despite encyclicals there was probably more understanding of some other sexual practices, which were regarded by many Christians as more natural than in the past, and without the horror that was formerly professed about them. Sexual intercourse, within marriage, was no longer generally regarded as polluted or inferior; and if Christians now wished to regard their religion as hallowing material things, and sex in particular, and so return to their Hebrew ancestry, many of them were better equipped to do so than at any time since the first century.

[11] J. Dominian, *Proposals for a New Sexual Ethic*, 1977.

Chapter 11

MODERN INFLUENCES

This survey of some of the sexual attitudes and practices of major living religions hardly leads to agreed conclusions, and it may serve rather to mark differences. But in modern times all religions are subjected to new pressures which are sure to affect their understanding of sex. As most of these influences come from the West it is Christianity and Judaism which have been the most changed so far, but other religions will be touched by acceptance or even by rejection.

MEDICINE

Some of these influences come from major developments in the following fields: medicine, psychology, women's rights, and the comparative study of religions. Medical knowledge has been revolutionary in providing more accurate understanding of the workings and results of sexual intercourse. Ancient western notions were based upon Greek philosophy and medicine which had no knowledge of the process of conception, and it was thought that the embryo was made from a mixture of semen and menstrual blood. Woman's role was simply to receive the precious semen, and this added to her subordination because she was regarded as a mere incubator, or a 'furrow' as Muslims said, and the man had 'the use of the woman'.

Until the sixteenth century the male semen was thought to be 'almost human', but then the essential female role was realized by the discovery of ovulation. Previously there had been horror at the 'waste' of semen, in masturbation, night 'pollution', and male homosexuality. Further, the 'bestial movements' involved in sexual intercourse had been thought by the church fathers to be the result of the Fall, since they were not under the control of the rational will. For a long time theologians ignored or refused to accept that coitus requires sensual excitement, and that orgasm is not a contradiction of the will but the expression of both will and energy. It is said that even in Victorian times women were told to shut their eyes and let the man express his lower nature, with the result that some men went to prostitutes for sexual pleasure and understanding.

Similar ignorance of the physiology of sex prevailed right across the world. For example, ancient China was also unaware of the fertilization which is effected by union of the male sperm cells with the female ova. All vaginal secretions were believed to constitute the Yin essence, and a principal aim of coitus was to enable the man to absorb the woman's Yin. The Chinese thought also that male semen was strictly limited, whereas women had an inexhaustible Yin essence. Every emission of semen was thought to diminish a man's Yang, which could only be compensated by gaining an equivalent amount of Yin from a woman, any woman. Ejaculation was often prevented either by mental discipline or physical pressure, and then the Yang fortified by Yin was thought to flow up the spinal column to strengthen the brain and the whole body and personality.

A similar belief was held by Indian Yoga, which may have derived it from China, but modern physiology regards such notions as erroneous, for if seminal fluid is checked it will enter the bladder and be discharged with the urine. According to ancient ideas semen was so valuable that both male masturbation and homosexuality were disapproved, though similar female practices were tolerated. Even involuntary nocturnal emissions were viewed with concern, since they only weakened the Yang, and might indeed be thought to be caused by demons or succubi stealing a man's vital essence by intercourse with him in dreams. Dreaming of a woman might show that she was an evil spirit, to be avoided if the man saw her in waking life. Similar notions were held in medieval Europe, where inquisitors earnestly debated whether witches or succubi could deprive men of their virile organ, and whether nuns were seduced by male incubi. The Chinese idea of the union of woman and man as Yin and Yang may have much to commend it, but the details need to be corrected in the light of modern physiology.

PSYCHOLOGY

Not only physiology but psychological investigations have revolutionized attitudes towards sex in the western world, and no religious interpretations should ignore them. The pioneer work of Freud revealed some of the sexual determinants of conduct and attitudes, factors that influence the emotional development of children, and the connections between neurosis and sexual repression. Then Jung's theory of the feminine or *anima* in man, and the masculine or *animus* in woman, contributed to further understanding of the complexity of sexual feelings. The study of dreams, which had long been held to be significant, though often interpreted in fantastic ways, threw further light on sexual desires.

In recent times case studies, such as the Kinsey and other reports, have provided masses of material for estimating patterns of desire and behaviour in men and women. Not only 'normal' sex, but homosexuality and practices which formerly were regarded as 'unnatural' are being viewed in new light. The large *Oxford English Dictionary* which formerly tersely defined masturbation as 'the action or practice of self-abuse', substituted new definitions in its 1976 *Supplement* as 'production of an orgasm by stimulation of the genitals, not by sexual intercourse; deliberate erotic self-stimulation'.

Further, the work of anthropologists among 'primitive' peoples has revealed new attitudes towards sexual intercourse, and varying roles played by men and women in different societies. There has also been a vast amount of literature on sexual subjects, though descriptions have been imaginary and often fantastic. D. H. Lawrence's *Lady Chatterley's Lover* was one of the most notorious and it was long banned in the land of its composition. While artistically probably not the best of his books, it has several curious features. Though written by a man it purports to give the woman's viewpoint in sexual intercourse, with what success no doubt women would be the best judges. In his later introduction or *Apropos* Lawrence wrote of 'the evocative power of the so-called obscene words', and claimed that we today are 'evolved and cultured far beyond the taboos which are inherent in our culture'. But it does not seem that his use of the tabooed words has made them much more acceptable in public usage. Yet most strange was his idea of the sexual attitudes of the Church. 'The Catholic Church', he claimed, 'especially in the south, is neither anti-sexual, like the northern churches, nor a-sexual like Mr Shaw and such social thinkers.' And he went back behind the Church to primitive actions which were claimed to fit in with the rhythm of the seasons. 'Sex goes through the rhythm of the year, in man and woman, ceaselessly changing: the rhythm of the sun in his relation to the earth', and so on with imagination but less attention to the complexity of human sexual behaviour.

The modern study of psychology has helped in the interpretation of symbols of waking life, as well as those of fantasy and dreaming. While phallicism should not be seen everywhere, its prevalence is of great significance and requires recognition and understanding. The 'serpent power' of Yogic theory resembled male sexual power rising up to unite with the highest psychic centre, referred to as a lotus which symbolized the vulva. The union of *linga* and *yoni* was depicted in imagery and painting in the cults of Shiva and Shakti, and formalized in yantric diagrams. Such patterns, as well as more obvious imagery, can be read and appreciated with the help of modern psychology, and the same may be said of pictures in many other parts of the world. From Freud and others modern people have learnt to interpret much of the unconscious symbolism of life and literature, from the Floating World to *Alice in Wonderland*. Medicine and psychology both help towards an understanding of physical intercourse and the involvement of the full personality in sexual relations.

WOMEN'S RIGHTS

It is apparent that in all living religions, and many ancient and prehistoric religions, men have been dominant, because physically stronger, and women have been subordinate. Whether women were the first to perform religious rituals, with societies organized on matriarchal bases, has been long debated, with much enthusiasm but little certainty. In historic times goddesses and female cults have abounded, but men have long dominated.

In ancient Greek society the law treated women as mere mechanisms for generating children, and such an attitude persisted in sexual teachings in many countries. Aristotle admitted that women had a deliberative faculty, but he said that it was 'without authority', and for him the male was the norm while women were deviant and inferior. The early Romans repressed women, though there were republican heroines who on occasion acted in opposition to men.

In most countries probably many women were respected and even loved by their menfolk, and verses were composed in their praise, like the virtuous wife of Proverbs, though sometimes these sound condescending. Upper-class women had more scope than the poor for exercising their special gifts, like Murasaki Shikibu in Japan, but often the most emancipated women were courtesans and geishas, who generally depended for their livelihood upon the favours of many men to whose desires they ministered.

The subordination of women was a prominent feature of Roman law and Jewish custom, reflecting male belief in the guarantee of positions of privilege by ancient tradition and divine appointment. Although it has often been claimed that Christianity gave a new status to women, in fact her social and legal position remained virtually unchanged. Paul had said that 'in Christ there is neither male nor female', but he also said that 'the head of every man is Christ, and the head of the woman is the man'. Believing women were sisters in the faith, but the force of ascetic tradition helped to keep women down. Chrysostom in the fourth century declared that the wife was not equal to her husband but must obey him, and even Clement of Alexandria earlier, while admitting that woman had an equal nature with man, also believed that man was better at everything. Misogynists like Tertullian and Jerome attacked women as the followers of Eve, who had first eaten of the forbidden fruit and so destroyed man. Fear was often expressed of women, who were like other Eves, always seeking to ensnare men like the foolish woman of Proverbs whose 'guests are in the depths of hell'. Such attitudes were disastrous for balanced sexual life, since woman was thought to have a demonic quality in her sexuality, and celibate priests and confessors could not easily give balanced advice on marital relations.

Despite their inferior position, in religious authority and usually in sexual life, women have formed the majority of lay followers of religion. Early Christianity owed much to its women, many of whom were named in the New Testament. Later outstanding women received honour: those who were martyrs for the Christian faith, and nuns and holy women in several religions. But even the most active were usually subordinate to men and they could not celebrate the sacraments. In some ancient religions, as in parts of modern Africa, there have been priestesses, but in most religions the highest ordination has been denied to women. It is remarkable that when state laws decree equal opportunities and employment for women, large religious bodies refuse them priestly ordination yet claim to have improved female status. Underlying some of the opposition to female ordination there may lie primitive fears of pollution by women's blood at the altar.

In Muslim countries the veiling of women may have sought to ensure their removal from temptation, as was claimed, but too often it resulted in their repression and seclusion, as part of a man's property, and men ran public affairs on their own. Polygamy, declared Edward Westermarck, in his *History of Human Marriage*, 'implies a violation of woman's feelings', though perhaps he was on less sure ground when he maintained that polyandry 'presupposes an abnormally feeble disposition to jealousy'.[1] In modern times the large part played by women

[1] 1891, pp. 495, 515.

in movements for national independence, notably in Gandhi's India, was a major factor enabling them to make further advances in social liberation, though there are still many obstacles to overcome. In many Asian and African countries, from Japan to Morocco, women are still almost entirely dependent upon their fathers or husbands for most things, and there are reactions towards traditional ways as well as advances.

The emancipation of women, especially in western countries, and therefore first affecting Christianity and Judaism, has been the outcome of many forces: urban, industrial, educational, and political. Much greater studies of and publicity about sexual matters, have enabled women as well as men to pronounce upon sexual questions, though publicity has also allowed vast commercial interests to profit by the exploitation of sex, and the degradation of women in pornographic literature and films. In such a situation, when old customs have been upset and new standards are hardly established, there is widespread confusion about the relations of sex and religion, or rather about the sexual attitudes and actions of religious people.

RELIGIOUS ENCOUNTER

The meeting of world religions is a further important feature of our times, and distinguishes them from previous ages when some religions were largely isolated from others, especially in the western world. In the nineteenth and twentieth centuries imperialism and missions took Christian ideas across Asia and Africa, and now immigration and travelling gurus return the compliment. Such contacts are as significant in the realm of sexual behaviour as anywhere else. The puritanism of the missions is now met alongside X films, and the ethereal Gita circulates with the Kama Sutra.

It is astonishing that nowadays the western world should often be regarded by much of the rest of mankind as both Christian and sexually loose and immoral. As we have seen, Christianity, like Buddhism, developed as an ascetic religion. In its regulations for clergy and monks, and the depreciation of sex in general, the Christian Church even surpassed

Buddhism. But 'nous avons changé tout cela', even those of us who remain practising Christians.

This picture of the West as sexually loose has been current for a long time, and it surprised people even in cultures where sexual practices were supposed to be much more naturalistic than in Christianity. Over twenty years ago in Omdurman, I observed closely-veiled women giving furtive glances at a large cinema poster which portrayed an American woman in tight jumper, breeches, and riding boots, standing whip in hand over the prostrate body of a man. In India, land of *lingas* and erotic cults, no kissing was allowed on cinema screens until recently. But now, in Asia and Africa, nudist films are imported from the West, and men and women profess shock at immoral films and magazines, and flock to see and read them. That the Christian West is immoral is a common judgement, along with envy at its material progress, and the traditional asceticism of Christianity is ignored or seen as a lost battle of the missions.

Such sexual confusion is part of the West's own problems, partly a product of reaction against 'Victorianism', encouraged by powerful commercial interests. It is often said that this is a 'permissive age', which some welcome and others deplore, but it is not clear whether the only alternative is a 'repressive age', following the swings back and forth which have marked western history in recent centuries.

The meeting of the great religious traditions of the world may bring help as well as challenge. It is sometimes asked what we can learn from other religions, and one factor is the understanding of sex. The ideal monogamy and love of Christianity, the world-affirmation of Judaism and Islam, the delight in sensual intercourse of classical Hinduism, the correlation of female and male in Chinese traditions, all these may contribute to new sexual ethics and tempered by each other they could mark a real advance. Of course none of these traditions need to be adopted indiscriminately, they need reform and adjustment and, in particular, most of them need to give fuller place and more respect to women. But by study of the relations of sex with the ideals of religions, and purgation of practices degrading to individuals, new and positive appreciation of the values of sex and love may be evolved.

Dr Joseph Needham, the great authority on *Science and Civilisation in China*, has spoken of the need for a new theology of sexuality:

> I have long been profoundly convinced that one of the greatest mistakes of Christian thinking through the centuries has been that sharp separation so many theologians and spiritual guides have made between 'love carnal' and 'love seraphick'. There are really no sharp lines of distinction between 'sacred' and 'profane' love, between *eros*, *phileia*, and *agape*. I believe that this division was essentially a Manichean belief intruding into the Christian gospel . . . We do urgently need today a new theology of sexuality . . . because of the fundamental new knowledge which man has gained since the seventeenth century about the nature of generation, and the structure and functions of his own mind. [2]

[2] Address for Caius Chapel, Cambridge, 1976.

SELECT BIBLIOGRAPHY

Allison, P., *African Stone Sculpture*. Lund Humphries 1968.

Archer, W. G., *The Loves of Krishna*. Allen & Unwin 1957.

Aston, W. G., tr., *Nihongi*. Kegan Paul 1896.

Aston, W. G., *Shinto, the Way of the Gods*. Longmans 1905.

Bailey, D. S., *Homosexuality and the Western Christian Tradition*. Longmans 1955.

Bailey, D. S., *The Man–Woman Relation in Christian Thought*. Longmans 1959.

Basham, A. L., *The Wonder that was India*. Sidgwick & Jackson 1954.

Basham, A. L., ed., *A Cultural History of India*. Oxford 1975.

Bell, R., tr., *The Qur'ān*. T. & T. Clark 1939.

Benedict, R., *The Chrysanthemum and the Sword*. Tuttle 1954.

Bester, J., tr., *Utamaro*. Kodansha 1968.

Bettenson, H., tr., *Augustine, The City of God*. Penguin 1972.

Bhattacharya, D., *Love Songs of Vidyāpati*. Allen & Unwin 1963.

Blacker, C., *The Catalpa Bow*. Allen & Unwin 1975.

Boyce, M., *Zoroastrians*. Routledge 1979.

Brandel-Syrier, M., *Black Woman in Search of God*. Lutterworth 1962.

Brazell, K., tr., *The Confessions of Lady Nijō*. Owen 1975.

Buber, M., *I and Thou*, tr. R. G. Smith. T. & T. Clark 1937.

Bühler, G., tr., *The Laws of Manu*. Oxford 1886.

Buitenen, J. A. B. van, tr., *The Mahābhārata*. Chicago 1973– .

Bunnag, J., *Buddhist Monk, Buddhist Layman*. Cambridge 1973.

Burton, R., tr., *Ananga Ranga*. New York Medical Press 1964 edn.

Burton, R., tr., *The Kama Sutra*. Allen & Unwin 1965 edn.

Burton, R., tr., *The Perfumed Garden*, Spearman 1963 edn.

Busia, K. A., *Social Survey of Sekondi-Takoradi*. Crown Agents 1950.

Chang, J., *The Tao of Love*, Wildwood 1977.

Chaudhuri, N. C., *Hinduism*. Chatto & Windus 1979.

Cohen, A., *Everyman's Talmud*. Dent 1949 edn.

Cohen, J. M., tr., *The Life of Saint Teresa*. Penguin 1957.

Cole, W. G., *Sex in Christianity and Psychoanalysis*. O.U.P. New York 1966.

Cole, W. O. and Sambhi, P. S., *The Sikhs*. Routledge 1978.

Comfort, A., tr., *The Koka Shastra*. Allen & Unwin 1964.

Cowell, E. B., tr., *The Jātaka*. Cambridge 1895.

Dale, K. J., *Circle of Harmony*. California 1975.

Daniel, N., *Islam and the West*. Edinburgh 1960.

Das, F. Hauswirth, *Purdah*, Routledge 1932.

De Bary, W. T., tr., *Five Women who loved Love*. Tuttle 1956.

De Bary, W. T., ed., *Sources of Indian Tradition*. Columbia 1958.

De Bary, W. T., ed., *Sources of Chinese Tradition*. Columbia 1960.

Dhavamony, M., *Love of God according to Śaiva Siddhanta*. Oxford 1971.

Dominian, J., *Proposals for a New Sexual Ethic*. Darton 1977.

Donaldson, D. M., *Studies in Muslim Ethics*. S.P.C.K. 1953.

Eliade, M., *Patterns in Comparative Religion*. Sheed & Ward 1958.

Eliade, M., *Yoga, Immortality and Freedom*. Routledge 1958.

Elwin, V., *The Religion of an Indian Tribe*. Oxford 1955.

Elwin, V., *The Tribal World of Verrier Elwin*. Oxford 1964.

Embree, J. F., *A Japanese Village*. Routledge 1946.

Epstein, I., *Judaism*. Penguin 1959.

Evans-Wentz, W. Y., *Tibetan Yoga and Secret Doctrines*. Oxford 1958.

Field, M. J., *Religion and Medicine of the Gā People*. Oxford 1937.

Forde, D., ed., *African Worlds*. Oxford 1954.

Fürer-Haimendorf, C. von, *The Naked Nagas*. Thacker 1962.

Goldziher, I., tr. S. M. Stern, *Muslim Studies*. Allen & Unwin vol. ii 1971.

Greer, G., *The Female Eunuch*. Paladin 1971.

Griaule, M., *Conversations with Ogotemmêli*. Oxford 1965.

Griffith, R. T. H., tr., *The Rig Veda*. Benares 1896.

Gulik, R. H. van, *Sexual Life in Ancient China*. Brill 1961.

Hamada, K., tr., *The Life of an Amorous Man*. Tuttle 1963.

Hastings, A., *Christian Marriage in Africa*. S.P.C.K. 1973.

Herbert, J., *Shintô*. Allen & Unwin 1967.

Hite, S., *The Hite Report*. Hamlyn 1977.

Horner, I. B., *Women under Primitive Buddhism*. Routledge 1930.

Horner, I. B., tr., *The Book of the Discipline*. Luzac 1938–52.

Hume, R. E., tr., *The Thirteen Principal Upanishads*. 2nd edn Oxford 1931.

Illing, R., *Japanese Erotic Art*. Thames & Hudson 1978.

Jacobi, H., tr., *Jaina Sūtras*. Oxford 1884.

Jacobs, L., *Hasidic Prayer*. Routledge 1972.

Jessop, T. E., *Law and Love*. Epworth 1948.

Jung, C. G., *Memories, Dreams, Reflections*. New York 1961.

Keyt, G., tr., *Gīta Govinda*. Bombay 1947.

Kidder, E., *Ancient Japan*, Elsevier 1977.

Kingsley, M. H., *Travels in West Africa*. Macmillan 1897.

Kinsey, A. C., *Sexual Behaviour in the Human Male*. Philadelphia 1948.

Kinsey, A. C., *Sexual Behaviour in the Human Female*. Philadelphia 1953.

Kisembo, B., *ed.*, *African Christian Marriage*. Chapman 1977.

Lebra, J., ed., *Women in Changing Japan*. Stanford 1976.

McIlvenna, T., *Meditations on the Gift of Sexuality*. Specific Press 1977.

Macquarrie, J., ed., *A Dictionary of Christian Ethics*. SCM 1967.

Maini, D. S., ed., *Sikhism*. Patiala 1969.

Marshall, P. J., ed., *The British Discovery of Hinduism in the Eighteenth Century*. Cambridge 1970.

Masters, W., and Johnson, V., *Human Sexual Response*. Boston 1966.

Mathers, P., tr., *The Book of the Thousand Nights and One Night*. 3rd edn Routledge 1973.

Mattuck, I., *Jewish Ethics*. Hutchinson 1953.

Meyer, J. J., *Sexual Life in Ancient India*. Routledge 1952.

Modi, J. J., *The Religious Ceremonies and Customs of the Parsees*. 2nd edn Karani 1937.

Morris, D., *The Naked Ape*. Cape 1967.

Morris, I., tr., *The Pillow Book of Sei Shōnagon*. Penguin 1967.

Murray, G., *Five Stages of Greek Religion*. Oxford 1925.

Needham, J., *Science and Civilisation in China*. Cambridge 1954– .

O'Flaherty, W. D., *Asceticism and Eroticism in the Mythology of Śiva*. Oxford 1973.

Pandey, R. B., *Hindu Samskāras*. 2nd edn Delhi 1969.

Parrinder, E. G., ed., *Man and his Gods*. Hamlyn 1971.

Parrinder, E. G., *Religion in an African City*. Oxford 1953.

Parrinder, E. G., *Mysticism in the World's Religions*. Sheldon 1976.

Philippi, D. L., tr., *Kojiki*. Tokyo 1968.

Rachewiltz, B. de, *Black Eros*. Allen & Unwin 1964.

Rattray, R. S., *Ashanti*. Oxford 1923.

Rawson, P., ed., *Primitive Erotic Art*. Weidenfeld & Nicolson 1973.

Rawson, P., *The Art of Tantra*. Thames & Hudson 1973.

Rawson, P., *Erotic Art of the East*. Weidenfeld & Nicolson 1973.

Rawson, P., *Erotic Art of India*. Thames & Hudson 1977.

Rawson, P. & Legeza, L., *Tao*. Thames & Hudson 1973.

Reynolds, B., tr., *Dante, La Vita Nuova*. Penguin 1969.

Rhys Davids, T. W., tr., *Dialogues of the Buddha*. Luzac 1899.

Robson, J., tr., *Mishkat-ul-Masabih*. Lahore 1960.

Scholem, G. G., tr., *The Zohar*, selections. New York 1949.

Scholem, G. G., *Major Trends in Jewish Mysticism*. Thames & Hudson 1955.

Seidensticker, E. G., tr., *The Makioka Sisters*. Secker & Warburg 1958.

Seidensticker, E. G., tr., *The Tale of Genji*. Secker & Warburg 1976.

Singer, M., ed., *Krishna; Myths, Rites, and Attitudes*. Chicago 1968.

Singh, T., tr., ed., Selections from *The Sacred Writings of the Sikhs*. Allen & Unwin 1960.

Smith, E. W., & Dale A. M., *The Ila-speaking Peoples of Northern Rhodesia*. Macmillan 1920.

Smith, M., *Rābi'a the Mystic*. Cambridge 1928.

Spiro, M. E., *Buddhism and Society*. Allen & Unwin 1971.

Stevenson, S., *The Heart of Jainism*. Oxford 1915.

Stevenson, S., *The Rites of the Twice-born*. Oxford 1920.

Thielecke, H., *The Ethics of Sex*. Clarke 1964.

Thompson, E., *Suttee*. Allen & Unwin 1928.

Thurian, M., *Marriage and Celibacy*. S.C.M. 1959.

Tsunoda, R., ed., *Sources of Japanese Tradition*. Columbia 1958.

Turner, V. W., *The Drums of Affliction*. Oxford 1968.

Vaudeville, C., *Kabīr*. Oxford 1974.

Vesey-Fitzgerald, S., *Muhammadan Law*. Oxford 1931.

Vidyarthi, L. P., ed., *Aspects of Religion in Indian Society*. Meerut 1961.

Waley, A., tr., *The Way and its Power*. Allen & Unwin 1934.

Waley, A., tr., *The Analects of Confucius*. Allen & Unwin 1938.

Watt, W. M., *Muhammad at Medina*. Oxford 1956.

Watts, A., *The Temple of Konarak*. Thames & Hudson 1971.

Webb, P., *The Erotic Arts*. Secker & Warburg 1975.

Welch, H., *The Parting of the Way*. Methuen 1957.

West, D. J., *Homosexuality*. Duckworth 1955.

Westermarck, E., *The History of Human Marriage*. Macmillan 1891.

Wilhelm, R., & Baynes, C. F., tr., *The I Ching*. Routledge 1951.

Woodroffe, J., *Principles of Tantra*. 4th edn Madras 1969.

Zaehner, R. C., *Mysticism Sacred and Profane*. Oxford 1957.

Zaehner, R. C., *The Dawn and Twilight of Zoroastrianism*. Weidenfeld & Nicolson 1961.

Zaehner, R. C. tr., *The Bhagavad-Gīta*. Oxford 1969.

Zaehner, R. C., ed., *The Concise Encyclopaedia of Living Faiths*. Hutchinson 1959.

Zaehner, R. C., *Concordant Discord*. Oxford 1970.

Zimmer, H., *Myths and Symbols in Indian Art and Civilization*. Bollingen 1946.

Zimmer, H., *Philosophies of India*. Bollingen 1951.

INDEX

———————

abortion 12, 47, 146, 169, 224, 237

adultery, Hindu laws 20; Jain 62; Parsi 71; tribal 73; African 143; Islamic 159, 163ff.; Hebrew 187; Christian 208ff.; 224 *and see* Fornication; Muhammad and Jesus on Mosaic law 164, 209

affinity 94, 145, 187 *and see* Exogamy, Incest, Taboos

Africa Ch. 7; Myths 128ff.; phallicism 131f.; initiation 134; marriage 137ff.; fertility 142f.; taboos 144f.; changes 146f.

agapé love 215ff., 248

Allison, P. 133, 249

Ama-terasu 104f., 117

Amida, Amitabha 96f., 122

anal intercourse 100, 163, 225

Ananga Ranga 30f., 167

androgyny 8, 18, 70, 129

Antony 12, 221

Aquinas 225, 228, 234

Arabian Nights 166f., 169, 177

Archer, W. G. 10, 249

art, erotic, Hindu 31f.; Buddhist 57f.; Jain 63f.; Chinese 82f.; Japanese 122f.; African 132f.; Christian 226

asceticism, of Shiva 7; in Gita 11f.; power 21ff.; Buddhist Middle Way 41f.; Jain severity 59f.; Greek 204f.; Gospel 210; Pauline 213; medieval 218ff. *and see* Tapas. Anti-asceticism, Sikh 68; Parsi 70; Islamic 162; Hebrew 179; Reformation 230; modern 246f.

Aston, W. G. 104, 106f., 118, 249

Augustine 39, 206, 222f., 224f., 228

Ayesha 153, 161f., 165, 173, 175

Bailey, D. S. 2, 3, 219ff., 226, 249

Basham, A. L. 40, 249

Baynes, C. F. 81, 254

Bernard 190, 228

bestiality, ritual 15; condemned 47, 74, 77, 100, 186, 222, 224f.; Animal marriages 108ff.

Bhagavad Gita 9, 11f., 246

Bhattacharya, D. 3, 249

Bible 141, Chs. 9–10

bigamy *see* Polygamy

birth control *see* Contraception

Blacker, C. 109, 116, 249

Boccaccio 221, 226, 230

bodhisattva 51, 107

body and soul *see* Dualism

Book of Common Prayer 232f.

Book of Rites 90, 93

Brandel-Syrier, M. 148, 249

bride-price *see* Dowry

brothels 27, 121f., 123 *and see* Prostitution

Buber, M. 217, 249

Buddhism Chs. 3, 5, 6; Buddha's birth 41f.; Order 43f., 95, 112, 221; celibacy 44ff.; temptations 45; Tantra 49ff.; lay morality 53ff.; women 55f.; Chinese 95ff.; Japanese 111ff.; in West 205f.

buggery, definition 225, *and see* Homosexuality, Sodomy

Buitenen, J. A. B. van 12, 250

Bunnag, J. 48, 54, 250

Burnaby, J. 223

bushido, way of warrior 116, 121

Busia, K. A. 139f., 250

cabbalism 198f.

Calvin, J. 230f.

Carstairs, G. M. 37

castration 21, 176, 219

Cathari 207, 225

celibacy, occasional 12, 15, 191; Buddhist 43ff., 48, 85, 95; Jain 61, 206; Christian 210, 212, 219ff., 235; rejected by Sikhs 68; Parsis 70; Chinese 95; Islam 162; Judaism 179, 191; Reformers 230f.

Chaitanya 32

chastity, Sita's 14; ascetics tempted 21ff.; Gandhi's 39; vow of 45, 63, 230; Buddhist 55; Parsi 71; Confucian women 90; Samurai women 116; Augustine deferring 222 *and see* Virginity

Chaudhuri, N. 10, 250

child marriage 16, 54

China Ch. 5, female and male 3, 77f.; Yin and Yang 79ff.; Supreme Ultimate Circle 80, 111; Tao 81ff.; Tao in sex 84ff.; manuals 86ff; Confucian morality 89ff.; marriage 93ff.; Buddhism 95ff.; variations 98f.; reactions 101f.

Christianity Ch. 10, Hebrew background 202, 207, 211; Greek 202ff.; Jesus 207ff.; Paul 211ff.; ascetic church 210, 213, 218ff., 230; Virgin Birth and Mother 226ff.; Reformation 229ff.; modern trends 235ff., 246f.; missions 39, 75f., 134f. 140, 147ff.; marriage depreciated 212, 219ff., 230; marriage sacrament 192, 224; monogamy 202, 215f.; on circumcision 135, 183f.; on Song of Songs 190; love 215f.

circumcision, African 130, 134ff.; Islamic 135, 160f.; Hebrew 181ff; not European, Indian or Chinese 181, 183; female *see* Clitoridectomy

Clement 206, 219, 244

clitoridectomy, African myth 130; practice 137, 147; Islamic 161, 177; not Hebrew 182

Cohen, J. M. 218, 250

coitus interruptus 163, 188

coitus reservatus, Tantric 38, 52; Chinese 85, 87, 88f., 93, 97, 241; Japanese 114, 123; African 139

Cole, W. O. 69, 250
concubines, Hindu 32f.; Sikh 68; Chinese 85, 93, 102; Japanese 115, 120; Islamic 151, 155, 175f.; Hebrew 193; Christian 220f., 230
Confucius, Confucians 52, 78, 85, 89ff.
contraception 18, 38, 163, 236f. *and see* Coitus interruptus and reservatus
courtesans 26, 29, 56, 72, 92, 94, 123, 204 *and see* Geishas, Hetairae, Prostitution
Cranmer, T. 231f.
cunnilingus 100, *and see* Oral sex

Dale, A. M. 127, 250
Dale, K. J. 126, 250
Daniel, N. 152, 250
Dante 4, 216
defloration 27, 53, 94, 170 *and see* Virginity
devadasi, temple servant 6, 27f.
dharma, duty, virtue 11ff., 24, 28
Dhavamony, M. 8, 250
divorce, of wife 71, 94, 139, 158f., 194, 207; of husband 71, 94; forbidden 207ff., 237; exceptions 206, 208, 237
Dominian, J. 239, 250
dowry, Hindu 16; Buddhist 54; Chinese 94; Japanese 119; African 137ff.; Islamic 155f.; Hebrew 194
dualism, body and soul, Indian 21ff.; Greek 204ff.; Manichee 206; Pauline 214; medieval 218ff; in Reformers 230ff.; Spirit and Nature 35; Heaven and Earth 77f., 151; good and evil 70; Yang and Yin 79ff., 97, 112f.; male and female 35f., 50f., 77ff.
Dubois, J. A. 39

Elwin, V. 74, 76, 250
Embree, J. F. 125, 251
Epstein, I. 201, 251
eros, love, passion 215ff., 248
Essenes 191, 202, 210
eunuchs 90, 166, 176, 208, 219
Evans-Wentz, W. Y. 52, 251
exogamy 74, 146

Fall of man 179, 222, 231 *and see* Original Sin
fellatio 100, 225 *and see* Oral sex
fertility 6, 71, 106f., 142, 190
Field, M. J. 143, 251
Fitzgerald, E. 172
Five M's *see* M
Floating World 121ff., 243
foot-binding 95, 101
fornication, penalties, Islamic 159, 163f.; Hebrew 186f.; Christian 208, 224f., 232 *and see* Adultery
Freud, S. 49, 108, 218, 221, 242f.
Fürer-Haimendorf, C. von 73f. 251

Gandhi, M. K. 39, 246
geisha 26, 121ff., 244 *and see* Courtesan, Prostitution
Genji, Tale of 114f., 121
Girl, Dark, Elected, Plain 86ff., 113f.
Gita Govinda 31f.
Gnostics 206f.
Goldziher, I. 174, 251
Greece, ancient 202ff., 240
Griaule, M. 131, 133, 251
Gulik, R. H. van 3, Ch. 5, 112, 114, 251

Hadith *see* Traditions

hair 67f.

harem 30, 33, 166, 175ff. *and see* Purdah, Zenana

harlotry *see* Prostitution

Hasidism 200f.

Hebrew *see* Judaism

Herbert, J. 106, 111, 251

hetairae 26, 204, 251 *and see* Courtesans, Geishas

Hinduism Ch. 2, sex of gods 5ff.; heroic ideals 10ff.; ritual union 17f.; marriage 15ff.; asceticism 21ff.; woman's lot 23ff.; prostitution 26ff.; sex manuals 28ff., 167, 246; erotic art 33ff.; Yoga 34f.; Tantra 35f.; puritanism 38f.

Hite, S. 161, 251

homosexuality, female, Hindu penalties 20; Chinese toleration 99; Pauline condemnation 186, 213; medieval attitudes 225 *and see* Lesbianism

homosexuality, male, Hindu toleration 21; Chinese 99; Japanese 122; African 146; Islamic condemnation 159, 163 and practices 169, 177; Hebrew penalties 186; Christian 213, 224, 238; Greek practices 204 *and see* Anal intercourse, Buggery, Sodomy

Horner, I. B. 46, 53, 57, 251

horse-sacrifice 15

Hosea 196, 211

houris 170

Hubbards 147

Humanae Vitae, encyclical 247

I Ching 80ff.

Immaculate Conception 228f.

impotence 22, 72, 169

incest, condemned, India 21; tribal 74; China 100; Japan 104; Africa 145f.; Islam 157; Hebrew 187; Christian 224f. *and see* Affinity

incubus 242

infanticide 54, 69, 173, 219, 224

infibulation 161

intercourse, definition 2 *and see* Sex

In, Yo 110ff.

Islam Ch. 8, monotheism 151, 195; Muhammad's marriages 151ff., polygamy 155f., 193f.; marriage 155ff.; Traditions 160ff.; sex literature 165ff.; symbolism 169ff.; women in early Islam 172f.; veiling 174f.; harems 175f.; circumcision 160f., 177

Izanagi, Izanami 103ff.

Jacobs, L. 200, 251

Jainism 21, 38, 41, 59ff., 205f.

Jamaa, family 136, 149f.

Japan Ch. 6, Shinto myths 103f.; phallicism 104, 106ff.; animal marriages 108f.; In and Yo 110f.; Buddhism 111f.; Tantra 112f.; sex manuals 101, 113, 121f.; women 114ff., 120, 123ff.; marriage 118ff.; Floating World 121ff.; geishas 124f.; new religions 125f.

Jerome 206, 220, 221, 231, 234, 245

Jesus 149, 207ff., 224, 227

Jizo 107, 119

Judaism, Hebraism Ch. 9, monotheism 178, 195f.; world-affirmation 189, 247; creation 178ff.; phallicism 180;

circumcision 181ff.; patriarchal 184f., 195; woman
184ff.; sex laws 186ff.; Song
of Songs 189f.; against celibacy 179, 191; marriage
191ff.; polygamy 193f.; symbolism 195ff.; influence on
Christianity 202, 207, 211,
236, 239, 247
Jung, C. G. 34, 229, 242, 251

Kabir 65f.
Kali 6, 52, 67
kama, desire, love 7, 8, 11, 13f.,
28f., 43
Kama Sutra 27, 28ff., 167, 246
karma, deeds 57, 59f., 122
Keyt, G. 31, 33, 251
Kidder, E. 107, 251
Kingsley, M. H. 147, 251
Kinsey, A. C. 242, 251
Kisembo, B. 141, 251
kissing, Indian 10, 29, 30, 32;
Chinese 87, 99f.; Islamic 161;
Hebrew 190; Aquinas on
danger 225; Indian cinema
taboos 247 *and see* Love
Koka Shastra 30f., 33
Kojiki 103f., 108f., 110, 117
Koran *see* Qur'an
Kripalani, K. 40
Krishna and Radha 9ff., 28ff.,
31, 67, 70, 189
kundalini, serpent power 35f.,
52
Kwanyin, Kannon 96

Lambeth Conferences 236f.
Lawrence, D. H. 242f.
lay morals, Hindu ideals 10ff.;
Buddhist rules 53ff.; Jain
vows 63; Sikh guide 69; Parsi
virtues 71f.; tribal taboos
74f.; Confucian morals 89ff.;

Japanese models 116f.;
Islamic traditions 161ff.;
Hebrew laws 186f.; Christian
rules 242ff. *and see* Marriage,
Sex
left-hand 36ff., 56, 86
Legeza, L. 82, 252
Lesbianism 20, 186 *and see*
Homosexuality, Female
Levirate 187
linga, phallus 5, 7f., 27, 33f.,
36f., 38, 47, 64, 75, 243 *and
see* Phallus
lotus, symbol 35, 97, 243
love, definitions 2, 215ff., 248;
Hindu ideals 11f., 13f.;
Chinese 81; Jamaa 149;
Hebrew 189ff., 196; married
love 149, 202, 211, 215ff., 247;
romantic 33, 138, 215ff., 239;
mystical 171, 198ff., 215ff.
Luther, M. 141, 230f.

M's, five 37, 112
maithuna, intercourse 37
Mahabharata 7, 10ff., 22f., 24
Mahavira 41, 59ff.
Makioka Sisters 120
male dominance, general 244;
Hindu 18, 24f., 29; Buddhist
50, 55; Chinese 78, 90f.;
Japanese 114, 116f., 124f.;
African 137; Islamic 157f.;
Hebrew 184; Christian 211f.,
214, 226 *and see* Woman
mandala, circle 33, 50, 112
Manichee 3, 70, 202, 206f., 214,
222, 225
mantra, text 28, 38, 50
manuals of sex, Hindu 28ff.;
Tantric 35f., 38, 49f., 95;
Chinese 78, 86f., 95, 100f.;
Japanese 113f., 122f.; Islamic
168f.

marriage, Hindu 15f., 20f.; Buddhist 54f.; Jain 63; Sikh 69; Parsi 71; Chinese 93ff., 97; Japanese 109, 118ff.; African 137ff.; Islamic 155ff., 162f.; Hebrew 187, 191f.; Roman 205; Christian 219f., 223ff.; Reformation 230ff.; 'spiritual marriage' 220; sacramental 167, 192, 224f.; mystical 97, 196, 211, 217; below virginity 213, 219f.; above virginity 219f., 230f., 237f.; made in heaven 192, 217 *and see* Child and Widow Marriage

Marshall, P. J. 39, 252

Mary, mother of Jesus 149, 171, 210, 219, 227ff.

Masters, W. and Johnson, V. 161, 252

masturbation, female, tolerated 99, 225, 241; forbidden 225

masturbation, male, forbidden in Buddhism 46; China 98; Christianity 189, 225, 238, 240; tolerated in Africa 146; modern 238f.; not in Bible 189, 238

matriarchy 145

menstruation, intercourse forbidden during, Hindu 18, 20; Parsi 72; tribal 74; China 86; Japan 117, 119; Africa 136, 145; Islam 157, 162; Hebrew 187f.; Christian 225; encouraged in Tantra 37; permitted by Taylor 234; danger to priests 75, 220, 245

Mercier P. 130

Meyer, J. J. 15, 252

modern changes Ch. 11

Modi, J. J. 71ff., 252

Mohammed *see* Muhammad

monasticism, monks, Buddhist 43ff., 95, 122; Jain 61f.; Christian 221ff., 230f.; rejected by Parsis 70f.; Islam 162; Reformation 230f. *and see* Nuns, Celibacy

monogamy, in Qur'an 156; in Genesis 193; Christian 140f., 207, 215ff., 247

morality *see* Lay morals

Mother Goddess 6, 9, 50, 73, 78, 129, 142, 195f., 229

Muhammad Ch. 8, 209

Murasaki Shikibu 114f., 121

Murray, G. 203, 252

Muslim *see* Islam

mut'a, temporary marriage 158, 174

mystical union, Indian 9f., 31f., 34f., 38; Chinese 80f., 84; Islamic 171f.; Hebrew 189f., 196f.; Christian 211, 217f., 223f. *and see* Marriage

nakedness *see* Nudity

Nanak 65ff.

nautch girls 27

Needham, J. 86, 101f., 248, 252

nirvana 50, 60f., 90

nudity, rare Hindu 9, 23; Tantric 36f., 50f., 98, 112; Jain monkish 51; tribal 76; Chinese 95; Japanese 105, 124; African 144, 145f.; Islamic ban 162; Adam and Eve 179f.; Greek 183; Medieval 210; complexes 146f., 210, 235

nuns 46, 49, 57, 61, 96f., 217f *and see* Monasticism

O'Flaherty, W. D. 8, 252

Omar Khayyam 172

Onanism 187, 188f.

oral sex 20, 29, 34, 100, 225
Original Sin 104, 180, 222, 228f.
Orthodoxy, Eastern 220f., 238
Oxford English Dictionary 1, 189, 242

Pandey, R. B. 16, 252
Parrinder, E. G. 73, 143, 252
Parsis 70ff., 79
patriarchy 158, 184f., 195f. *and see* Male dominance
Paul of Tarsus, dualism 180, 214, 236; on marriage 211ff., 224, 234; on circumcision 183f.; on Lesbianism 186, 213; on Greek gods 203; on celibacy and virginity 212f., 219
peach, symbol 82f., 107, 180
penis 34, 130f., 169, 188 *and see* Phallus
penis-sheath 144
Perfumed Garden 168ff.
phallus, in India 5, 7f., 33, 39, 75, 243; Tantra 36f., 50, 97; China 82; Japan 106ff., 119; Africa 131ff.; Hebrew 186ff., 199; Greek 203; Europe 218, 226 *and see* Linga, Shiva
philia, friendship 215, 248
Philippi, D. L. 104, 108, 252
Plato 18, 70, 203f.
polyandry 15, 23, 158, 245
polygamy, polygyny, Hindu 30; Buddhist 55; Sikh 68; Parsi 72; Chinese 85, 90f.; African 140f., 145; Islamic 151, 155ff., 175f.; Hebrew 184f., 193; Christian attitudes 140f., 215, 231, 245 *and see* Monogamy, Polyandry
pornography 46, 100, 122, 142, 246 *and see* Art Manuals
possession 109, 118

procreation, primary purpose, India 18f.; China 85, 90, 100; Africa 146; Islam 156; Judaism 192; Christianity 223, 232; primacy questioned 236 *and see* Coitus reservatus
profane love 215, 248
prostitution, Indian 26f.; Buddhist 56f., 97; Chinese 97, 100; Japanese 121ff.; Muslim 177; Greek 204; Hebrew symbolism 189; toleration 225f.; condemnation 72, 186, 196, 208, 234 *and see* Courtesans, Geishas, Hetairae
purdah 63, 176
puritanism, rigorism, recent Indian 39f., 76; Manchu 101; Islamic 38, 160, 177; Roman 205; Christian 39, 127, 230, 235 *and see* Victorianism

Qur'an, Koran Ch. 8, 227

Radha 9f., 28ff. *and see* Krishna
Rama and Sita 14, 16, 26, 59
rape 20, 188, 203, 218, 224f.
Ratirahasya *see* Koka Shastra
Rattray, R. S. 131, 137, 252
Rawson, P. 82, 252
religion, definition 2, encounter 246f.
Robson, J. 161, 253
romance *see* Love

sacramental sex 31, 167f., 224
sacred love 215, 248
Sambhi, P. S. 69, 250
Samkhya 35ff.
Samurai 111, 116
sati, suttee, widow-burning 24f., 65, 68

Scholem, G. G. 199, 200, 253
semen, sacred, in India 18f.; Tantra 38, 52, 85, 97f.; China 85ff., 97ff., 241; Judaism 188; Christianity 225, 240 *and see* Coitus reservatus
seraphic love *see* sacred
sex, definitions 1 *and see* Love; intercourse 2; Indian ideas 17f., 29f.; Jain 62f.; Chinese 77ff.; Islamic 157f.; Hebrew 192f.; Christian 211f., 233f.; Tantric 37, 51; extra-marital 30, 73, 188, 238 *and see* Adultery, Chastity, Fornication, Prostitution. Divine sex, Hindu 7f., 9f., 28, 31ff.; Chinese 78f.; Japanese 103ff.; African 129f.; Hebrew thought 196ff.; Christian 211f., 217f. *and see* Mystical union
Shakti 9, 31f., 36f., 67
Shekhinah 199f.
Shinto Ch. 6
Shiva 6, 7, 16, 31, 33, 67, 70, 75, 151, 172, 203, 243
Sikhs 65ff.
sin *see* Original sin
Smith, E. W. 127, 144, 253
Smith, M. 173, 253
snake 35f., 108f., 130, 131, 180f.
Sodomy, definition 186, 222, 225f. *and see* Homosexuality, Male
Song of Songs 32, 189ff., 216
Spiro, M. 48, 253
Stern, S.M. 174, 253
Stevenson, S. 16, 17, 25f., 39, 63, 253
succubus 98, 241
suttee *see* Sati
swastika 64

taboo, Hindu 20, 37, 67; Buddhist 47, 50; Sikh 67; tribal 73f.; Chinese 94f.; Japanese 104; African 128, 144ff.; Hebrew 188; modern 243; broken taboos *see* Tantra
Tachikawa sect 112f.
t'ai chi t'u, circle 80, 111
Takasago couple 120
Talmud 184ff., 192ff., 198ff.
Tantra 9, 19f., 28, 34ff., 49ff., 64f., 86, 96f., 112f.
Tao 91ff., 84ff., 91ff.
Tao Te Ching 78, 81
tapas, asceticism 7, 8 *and see* Asceticism
tattoo 136ff.
Taylor, J. 233f.
tea ceremony 111
Teresa 218
Tertullian 219, 228, 245
Thompson, E. 25, 253
Traditions, Islamic 160ff.
transvestism 29, 32, 122, 142, 146, 186
Trent, Council 140, 223, 235
Tsunoda, Ryusaku 111, 116, 253
Turner, V. W. 136, 253
twins 130f., 133

Uberoi, J. P. S. 68
ukiyo, floating world 121ff.
Upanishads 3, 7, 17ff., 43

vagina *see* Vulva
Vaudeville, C. 66, 253
veiling of women 25, 69, 90, 174ff., 245
Vesey-Fitzgerald, S. 167, 253
Victorianism 39, 146f., 189, 218, 230, 235, 241, 247
Virgin Birth 41f., 206, 226ff.

Virgin Mary *see* Mary
virginity 23, 93, 163, 187f., 213, 219f., 233, 235 *and see* Celibacy, Chastity
vulva, vagina 6, 18, 82f., 130, 136, 243 *and see* Yoni, Clitoridectomy

Waley, A. 78, 90, 253
Watt, W. M. 152, 253
Watts, A. 34, 253
Welch, H. 82, 253
Westermarck, E. 245, 253
widow, burning *see* Sati
widow, marriage banned 17, 63, 95
Wilhelm, R. 81, 253
woman, danger of 44f., 62, 66, 74f., 220, 245; eroticism of 23, 36f., 50f., 77f., 86f., 113, 166, 169, 190 *and see* Tantra, Yin and Yang, Mother Goddess. Inferior status of woman in Hinduism 23ff.; Buddhism 55; Jainism 61f.; Kabiri 66; China 90, 92, 95, 102; Japan 114f., 116ff.; Islam 174ff., 240; Judaism 184ff., 244; Greece 240, 244; Rome 244; Christianity 211ff., 214, 229, 232, 244 *and*
see Male Dominance, Polygamy. Higher status of woman 55, 68, 77, 86, 102, 172f., 184, 205, 229, 244ff. *and see* Tantra
world-affirmation 39, 67, 70, 84, 92, 146ff., 151, 155, 178, 189, 235, 239, 247
world-denial 21, 43, 59f., 92, 205, 219ff., 230 *and see* Asceticism, Dualism

Yang and Yin 2, Ch. 5, 110ff., 151, 241
yantra, diagram 64, 243
Yellow Emperor, Huang Ti 86f., 113
Yin *see* Yang
Yoga 3, 6, 34ff., 43, 49f., 83
yoni, vulva 6, 8, 9, 17, 33, 47, 51f., 58, 64
Young, J. de 54

Zaehner, R. C. 39, 70, 102, 218, 233, 253
Zen 30, 176
zenana 30, 176 *and see* Harem
Zimmer, H. 17, 253
Zohar 198f.
Zoroaster, Zarathushtra 79 *and see* Parsis
Zwingli, U. 230